BestMasters

Springer awards „BestMasters" to the best master's theses which have been completed at renowned universities in Germany, Austria, and Switzerland.

The studies received highest marks and were recommended for publication by supervisors. They address current issues from various fields of research in natural sciences, psychology, technology, and economics.

The series addresses practitioners as well as scientists and, in particular, offers guidance for early stage researchers.

Florian Jacob

Risk Estimation on High Frequency Financial Data

Empirical Analysis of the DAX 30

 Springer Spektrum

Florian Jacob
Karlsruhe, Germany

BestMasters
ISBN 978-3-658-09388-4   ISBN 978-3-658-09389-1 (eBook)
DOI 10.1007/978-3-658-09389-1

Library of Congress Control Number: 2015936902

Springer Spektrum

Springer Spektrum is a brand of Springer Fachmedien Wiesbaden
Springer Fachmedien Wiesbaden is part of Springer Science+Business Media
(www.springer.com)

Acknowledgements

This work would not have been possible without the support, encouragement, and advice of several people. Foremost, I would like to thank Dr. habil. Young Shin (Aaron) Kim for the introduction to the topic during my time as a master student at the Karlsruher Institute of Technology, as well as for the guidance and support throughout this thesis. I would also want to express my gratitude to Prof. Dr. Wolf-Dieter Heller and Prof. Dr. Svetlozar (Zari) Rachev for giving me the opportunity to pursue my interests in Germany and during my two month stay at Stony Brook University, USA. This time as well as helping me to develop a professional skill set, let me develop as a person.

Florian Jacob

Contents

List of Figures

All figures can be accessed on **www.springer.com** under the author's name and the book title.

List of Tables

1. Introduction

"Wednesday is the type of day people will remember in quant-land for a very long time
...Events thats models only predicted would happen once in 10,000 years happened every day
for three days."

(The Wall Street Journal 2007)

For the last 10 years, we have witnessed a vast development of capital markets, introducing high frequency trading and a shift of market share towards high-frequency and algorithmic-trading (Itai and Sussmann 2009). In the recent past, high-frequency trading and automated trading has been accused for enforcing price shocks and rising volatility. Consequently it is of high interest in the academic world of understanding the complex phenomena and processes that dominate financial markets on short time horizons. To understand these phenomena is not just of special interest in the academia but also of particular interest in the financial world. Financial firms that trade assets on high-frequency time scales also seek to extend their knowledge about financial processes on short time intervals. Within their framework of high-frequency trading, also risk has to be estimated on a intraday basis.

The possibility of intraday analysis has only occured during the last few years[1], before analysis was performed on an end of day basis. In general, modern finance relies heavily on the assumption that the random variables under investigation follow a normal distribution. Distributional assumptions for financial processes have important theoretical implications, given that financial decisions are commonly based on expected returns and risk of alternative investment opportuinities. Hence, solutions to such problems like portfolio selection, option pricing and risk management depend critically on distributional specifications. By now there is however ample empirical evidence that financial return series deviate from the Gaussian model in that their marginal distributions are heavy-tailed and, possibly, asymmetric. The countermovement against the Gaussian assumption in financial modeling started as early as in the 1960s. Beginning with the work of Mandelbrot (1963),(1967) and Fama (1965). In their investigations the repeating occurence of violent fluctuations of observable prices for different tradable goods, such as cotton, led them to the conclusion of strictly rejecting the Normal assumption for price processes. Usually financial data is presented in terms of continously compounded log-returns $r_t = \log \frac{S_{t+\Delta t}}{S_t}$ derived from the series of corresponding stock prices S_t. We use the term stock and risky asset equivalently in the following and denote the stock price at time t by S_t. Financial data exhibits so-called stylized facts:

1. Heavy-Tailedness or Excess Kurtosis: Empirical return distributions have more pronounced tails in addition to a more peaked center compared to the Gaussian assumption.

2. Asymmetry or Skewness. Whilst normal distributions are always symmetric around their mean, observable returns mostly exhibit asymmetry in favour of large negative return deviations.

3. Volatility Clustering: In reality, large price movements in one period tend to be followed by equally large price movements in the next period. In the same manner, after calm periods one is more likely to observe only moderate price changes in the following time step.

4. Price Jumps: Financial tick data is available on a rather high-frequency scale today, but

---

[1]for differnet studies in high-frequency data see (Hendershott and Riordan 2012) and (Riordan and Storkenmaier 2012).

is still collected in discrete time. This makes it difficult to tell whether large and sudden moves in asset prices correspond to actual jumps in the chosen continuous time modelling or are merely due to discrete sampling. Nevertheless, it is proven evidently that empirical observations cannot be implied by diffusion processes with continuous trajectories alone, but necessarily require the incorporation of jumps in the employed underlying models.

Both asset management and option pricing models, require the proper modelling of the return distribution of financial assets. Therefore, an extensive application of the theory of stochastic processes to finance took part and the literature on asset and option pricing models grew rapidly over the last decades. There are mainly two ways that have been proposed in the literature to deal with non-normality. The first is to include stochastic volatility and allow the variance of the normal distribution to change over time. The model by Heston (1993) is the most well-known model using this approach. The other approach uses jumps in the return model and was first introduced by Merton (1976). In this master thesis we start with the second approach and introduce tempered stable distributions, which are capable of capturing both asymmetry and heavy tails. The form of tempered stable distribution studied in this master thesis is the normal tempered stable (NTS) distribution, that was first studied by Barndorff-Nielsen and Shephard (2001). The emphasis of this thesis lies on the NTS distribution, its multivariate extension and its appliciation in the context of different ARMA-GARCH models.

2. Theory of Time Series Modeling and Risk Estimation

"...as you well know, the biggest problems we now have with the whole evolution of risk is the fat-tail problem, which is really creating very large conceptual difficulties. Because as we all know, the assumption of normality enables us to drop off the huge amount of complexity in our equations... Because once you start putting in non-normality assumptions, which is unfortunately what characterizes the real world, then these issues become extremely difficult."

(Greenspan 2007)

2.1. Financial Econometrics

Forecasting the future behavior of stock prices is an essential task in the implementation of risk management systems. In order to obtain a good forecast for the distribution of returns, prediction of future market volatility is critical. A key modeling difficulty is that market volatility cannot be observed directly. Volatility musst be inferred by looking at the past behavior of market prices or at the value of financial derivatives. Appllying the first case, one can state that if prices fluctuate a lot, volatility should be high. The ascertainment of how high is difficult. One reason is that it can not be stated whether a large shock to prices is transitory or permanent. Due to the latent character of the variable a statistical model has to be applied, making strong assumptions . In the case of financial data the stylized facts of nonnoramlity , volatility clusters and structural breaks represent a demanding challenge in the process of modelling volatility. In the basic time series approach, ARMA models are used to model the conditional expectation of a process given the past, but in ARMA models the conditional variance given the past is constant. For example an ARMA model can not capture that if the recent returns have been unusually volatile, it might be expected that the upcoming return is also more variable. To solve the problem of non linearity in variance the ARCH model was proposed by Engle (1982). The characteristic of volatlity clustering is encompassed in the more general feature of *heteroskedasticity* in financial time series In order to allow for past conditional variances in the current conditional variance equation Bollerslev (1986) introduced the generalized ARCH (GARCH) model. Subsequently of this developement a rich family of GARCH models has emerged.

2.1.1. GARCH models

The ARCH(r) model, as introduced by (Engle 1982), describes the process y_t by

$$y_t = \epsilon_t \tag{2.1.1}$$

$$\epsilon_t = \sqrt{h_t}\eta_t, \quad \eta_t \overset{IID}{\sim} N(0,1) \tag{2.1.2}$$

$$h_t = a_0 + \sum_{i=1}^{q} a_i \epsilon_{t-i}^2 \tag{2.1.3}$$

where h_t is the variance of ϵ_t conditional on the information available at time t. h_t is called the *conditional variance* of ϵ_t. In order to have a well defined process, conditions on the coefficients need to be imposed to avoid negative h_t values. To ensure this the parameters of the model must satisty $a_0 > 0$ and $a_i \geq 0$ for $i = 1, 2, \ldots, q$. The random variable η_t is an innovation term which is assumed to be IID with mean zero and unit variance. The Gaussian assumption is not critical as it can be relaxed to allow for more heacy-tailed distributions[1]. Typical distributions are discussed later in the thesis.

(Bollerslev 1986) extended the ARCH model by appending a weighted sum over the past p conditional variances h_{t_i} for $i = 1, 2, \ldots, p$. The resulting model is known as *generalized autoregressive conditional heteroskedastic model*, GARCH-model. Hence,

$$\epsilon_t = \sqrt{h_t}\eta_t, \quad \eta_t \overset{IID}{\sim} N(0,1) \tag{2.1.4}$$

$$h_t = a_0 + \sum_{i=1}^{q} a_i \epsilon_{t-i}^2 + \sum_{j=1}^{p} b_j h_{t-j} \tag{2.1.5}$$

$$\tag{2.1.6}$$

where, again, h_t denotes the conditional variance of ϵ_t (conditional on the information available at time t).For the process to be well defined and to ensure that the conditional variance of ϵ_t is stationary within the GARCH-model framework, the parameters a_i and b_j must satisfy the necessary and sufficient conditions: $\sum_{i=1}^{q} a_i + \sum_{i=1}^{q} < 1$, $a_0 \geq 0$and $a_i > 0$, $b_j > 0 \forall i \in \{0, \ldots, q\}$ and $j \in \{0, \ldots, p\}^2$.

In order to apply a GARCH model to financial time series, it has to be ensured that the conditional return distributions are centered around zero. This can be achieved by de-meaning the complete time series or by applying a conditional mean model (e.g. an ARMA model).

---

[1]For a profound description of the porperties of the ARCH model we refer to Francq and Zakoian (2010), Rachev et al. (2007).

[2]For a profound description of the GARCH model and for proofs of the conditions we refer to Francq and Zakoian (2010).

2.1.2. FIGARCH

In order to accomodate empirical regularities concerning the evidence of fractional integration for a time-series of daily index returns (Baille et al. 1996) suggest the 'fractionally integrated GARCH' (FIGARCH) class to incorporate long-memory behavior in the squared return or absolute return process. The development of the FIGARCH model was a continuation of the development of GARCH (Bollerslev 1986) and IGARCH (Engle and Bollerslev 1986). The FIGARCH model is mostley inherited from the ARFIMA-class. The FIGARCH is to allow for fractional order of integration such that

$$(1 - a_1 L - b_1 L)\epsilon_t^2 = \phi(L)(1 - L)^d. \tag{2.1.7}$$

. A FIGARCH(p,d,q) process is defined by

$$a(L)(1 - L)^d \epsilon_t^2 = a_0 + b(L)u_t^2 \tag{2.1.8}$$

where $u_t = \epsilon_t^2 - h_t^2, a_0 \in (0, \infty)$, and $(1 - L)^d$ is a fractional difference operator. Substituting the innovations ϵ_t in 2.1.8, we get an expression for the conditional variance, namely

$$\sigma_t^2 = \frac{c_0}{1 - b(1)} + \underbrace{\left(1 - \frac{\phi(B)(1 - B)^d}{1 - b(B)}\right)}_{= \lambda(B)} r_t^2, \tag{2.1.9}$$

where $\lambda(B) = \lambda_1 B + \lambda_2 B^2 + \ldots$ is an infinite-order polynomial. Thus, the latter might be interpreted as an ARCH(∞)-representation of $r_t = \sigma\epsilon_t$ with conditional variance σ_t^2. In order to have a reasonable specification, the conditional variance σ_t^2 must be positive implying further constraints on the parameters of $\phi(B)$ and $b(B)$. The constraints are not trivial and there have been several attempts to concretize them, mostly case by case. Baille et al. (1996) examine the FIGARCH(1,d,0)-case and obtain for the coefficients λ_k of $\lambda(B)$,

$$\lambda_k = (1 - b_1 - (1 - d)k^{-1})\frac{\Gamma(k + d - 1))}{\Gamma(k)\Gamma(d)}, \tag{2.1.10}$$

which implies $0 \leq b_1 \leq d \leq 1$ as a sufficient condition for almost surely positive conditional covariance. Applying Stirling's formula to λ_k, 2.1.9 shows that shocks in the conditional variance decay hyperbolically and thus the FIGARCH(1,d,0) has some form of long-memory behavior. Chung (1999) gives an admisable range of FIGARCH(1,d,1)-processes, i.e. $0 \leq \phi_1 \leq b_1 \leq d \leq 1$, whereas Bollerslev and Mikkelsen (1996) state the constraints

$$b_1 - d \leq \phi_1 \leq \frac{2-d}{3} \text{ and } d(\phi_1 - \frac{1-d}{2}) \leq b_1(\phi_1 - b_1 + d).$$

Conrad and Haag (2006) the correctness of both parameter settings. In this thesis we use the FIGARCH(1,d,1)-process. For parameter constraints on the FIGARCH(1,d,q) and the FIGARCH(p,d,q) we refer to the paper of Conrad and Haag (2006).

2.2. Model Choice and Validation

This section considers the problem of selecting an appropriate model and introducing different methods to display the behavior of time series. A large part of the theory of finance rests on the assumption that prices follow a random walk. The price variation process, should thus consitute a martingale difference sequence (Francq and Zakoian 2010) and should coincide with its innovation process. A consequence of this is the absence of correlation. In order to detect autocorrelation we introduce the Sample Autocorrelation Function and the Sample Partial Autocorrelation Function. In case the assumption of no correlation cannot be sustained, ARMA models could be applied to model the data before using GARCH models for the residuals. As literature such as Baille et al. (1996) and Ding et al. (1993) suggests modeling the temporal dependence in volatility experiences long range dependency. Using measures of volatility such as powers or logarithms of squared returns, these authors have found that the sample autocorrelation function of volatility decays slower than exponentially. In order to further investigate the long-range dependence we estimate the self-similarity parameter H.

2.2.1. Detecting the Autocorrelation

Before we apply a certain time series model, it is reasonable to calculate the time series' autocorrelation and independence structure. In order to visually detect autocorrelation the Sample Autocorrelation Function (SACF) adn the Sample Partial Autocorrelation Function (SPACF) can be used. The SACF is defined as follows:

$$\hat{\rho}_k = corr(y_t, y_{t-k}) = \frac{cov(y_t, y_{t-k})}{\hat{\gamma}_0}, \qquad\qquad k = 0, 1, ... \qquad (2.2.1)$$

where $\hat{\gamma}_0$ is the unconditional variance of the time series: $\hat{\gamma}_0 = Var(y_t)$. $\hat{\rho}_k$ denotes the sample autocorrelation coefficient between y_t and y_{t-k}. The SPACF is construcetd as

$$\hat{\alpha}_k = corr(y_t, y_{t-k} | y_{t-1}, y_{t-k+1}). \qquad\qquad k = 1, 2, ... \qquad (2.2.2)$$

The SPACF describes the correlation between y_t and y_{t-k} after eliminating the linear dependence between the realizations y_{t-1} through y_{t-k+1}. When applying pure AR or MA models, these two methods can be directly used in order to find the appropriate number of model parameters. In more general cases, the SACF and SPACF are useful tools in order to detect non-stationarity and long-range-dependence (LRD) effects in time series. LRD is indicated if the SACF is declining slower than exponentially. One frequently employed measure is the SACF constructed from a time series of squared returns to investigate the time series for dependence. SACF values which are significantly different from zero indicate that the squared returns are correlated and suggests the application of GARCH models. If the SACF of squared returns exhibits persistent behavior, that is, the correlation between squared returns is not declining exponentially with increasing lag, long range dependency is indicated which motivates the utilization of FIGARCH models.

2.2.2. Testing for Long-Range Dependency

Many different methods are available for estimating the self-similarity parameter H. Some of which are described in detail in the monograph of Beran (1994). They are typically validated by appealing to self-similarity or to an asymptotic analysis where one supposes the sample size converges to infinity. In this thesis we use the better known R/S method [3]

We briefly discuss the R/S method as described in Taqqu et al. (1995). It is discussed in detail in Mandelbrot and Wallis (1969) and Mandelbrot and Taqqu (1979).

For a time series $\{X_t\}_{t \in T}, T = \{1, \ldots, N\}$, with partial sum $Y(n) = \sum_{i=1}^{n} X_i, n = 1, \ldots, N$, and sample variance $S^2(n) := (1/n) \sum_{i=1}^{n} X_i^2 - (1/n)^2 Y(n)^2$, the R/S statistic, or the *rescaled adjusted range*, is given by:

$$R/S(n) := \frac{1}{S(n)} \left[\max_{0 \leq t \leq n} \left(Y(t) - \frac{t}{n} Y(n) \right) - \min_{0 \leq t \leq n} \left(Y(t) - \frac{t}{n} Y(n) \right) \right]. \qquad (2.2.3)$$

The common proceeding to determine H using the R/S statistic is to subdivide a times series of length N, into K blocks, each of size N/K. Then, for each lag n, compute $R(k_i, n)/S(k_i, n)$, starting at points $k_i = iN/K + 1, i = 1, 2, \ldots$, such taht $k_i + n \leq N$. For values of n smaller than N/K, one gets K different estimates of $R(n)/S(n)$. For values of n approaching N, one gets fewer values, as few as 1 when $n \geq N - N/K$. Following plot $\log[R(k_i, n)/S(k_i, n)]$ versus $\log n$

---

[3]For a comparison of perfomance of different methods and as an overview of the existing methods we refer to Taqqu et al. (1995).

and get, for each n several points on the plot. The parameter H can be estimated by fitting a line to the points in the plot. Usually the high and low ends of the plot are cut off.

However, the estimate of the actual H from R/S-statistic should be handled with care and can be taken as an indicator for present long-memory at most. Beran (1994) mentions the difficulty to identify the distribution of $R(n)/S(n)$, as well as their dependencies on each other and therefore the lack of any confidence intervals for the estimates. He shows that the statistic is not robust against slowly decaying trends and other departures from stationarity in the data. On the pro side, the R/S-statistic has robustness against infinite variance of the data generating process as well as simple and fast computation, which makes the statistic a good candidate for an approximation of the actual long-memory behavior.

2.3. The innovation process

In order to derive the NTS Process, this section presents the fundamental theoretical building blocks. Therefore, we give a brief definition of Lévy Processes, infinitely divisibility, Subordination and show two general concepts of distributions the α-stable and the Classical Tempered Stable which is used for construction of the Normal Tempered Stable.

2.3.1. Lévy Processes

In probability theory a Lévy Process is a stochastic process with independent, stationary increments, named to the honour of French mathematician Paul Lévy. It represents the continuous-time analog of a random walk. Fundamental examples of the Lévy Process are Brownian Motion and Poisson Process. A cadlag[4] stochastic process $(X_t)_{t \geq 0}$ defined on the probability space $(\Omega, \mathcal{F}, \mathbb{P})$ with values in $\mathbb{R}^d$ such that $X_0 = 0$ is called a Lévy Process if it possesses the following properties:

1. Independent increments: for every increasing sequence of times $t_0, ..., t_n$, the random variables $X_{t_0}, X_{t_1} - X_{t_0}, ..., X_{t_n} - X_{t_{n-1}}$ are independent.

2. Stationary increments: the law of $X_{t+h} - X_t$ does not depend on t.

3. Stochastic continuity: $\forall \epsilon > 0, \lim_{h \to 0} \mathbb{P}(|X_{t+h} - X_t| \geq \epsilon) = 0$.

For an extensive overview over their theoretical and practical properties we refer to a variety of textbooks such as Sato (1999) or Kyprianou (2006). The link between mathematical formulation and arising economic modelling consequence presents itself as follows. Independent increments mean that we assume independence between past and future returns. Stationarity assumes that past and future returns are equally distributed. Both of these assumptions are violated to a certain degree by empirical evidence. As shown in the introduction the stylized fact of non-constant volatility is against the second property as well as the first is violated by existing autocorrelation of returns shown in Figure 2.1. A solution to these problems shall be given in Section 3.4.3.

---

[4]The class of processes with right continuous paths and left limits is referred to as cadlag process.

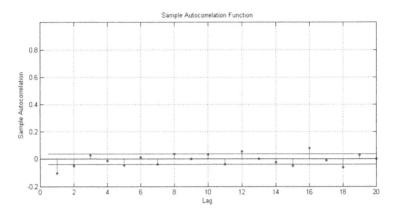

FIGURE 2.1.: *Illustration of autocorrelation of the S&P 500 of 01.01.2002 - 22.10.2012*

Definition 2.1 (Infinite divisibility).

A probability distribution F on $\mathbb{R}^d$ is said to be infinitly divisible if for any integer $n \geq 2$, there exist n independent and identical distributed (i.i.d.) random variables $Y_1, ..., Y_n$ such that $Y_1 + ... + Y_n$ has distribution F.

To clarify the connection and the use of the given definitions, the following lemma is given.

Lemma 2.3.1.1 (Correspondence Between Lévy Processes and Infinite Divisibility).

Let $(X_t)_{t\geq 0}$ be a Lévy process. Then for every t, X_t has an infinitely divisible distribution. Conversely, if F is an infinitely divisible distribution then there exists a Lévy Process (X_t) such that the distribution of X_1 is given by F.

For proof we refer to Sato (1999), section 11.6. For example the normal distribution is infinitely divisible and used in the famous Black-Scholes model for option pricing. Further Poisson, Gamma, Variance Gamma (VG), Inverse Gaussian (IG), α-stable, Classical Tempered Stable (CTS), General Tempered Stable (GTS), Normal Tempered Stable (NTS), Rapidly Decreasing Tempered Stable (RDTS) and Kim-Rachev Tempered Stable (KRTS) are infiinitly devisible

(Rachev et al. 2011) and thus can be used for the construction of Lévy Processes. The characteristic function of the one-dimensional infinitely divisible distribution is generalized by the Lévy-Khinchine formula:

$$\exp\left(i\gamma u - \frac{1}{2}\sigma^2 u^2 + \int\limits_{-\infty}^{+\infty} (e^{iux} - 1 - iux 1_{|x|\leq 1})\nu(dx)\right). \tag{2.3.1}$$

The measure ν is referred to as *Lévy measure*, representing the jump component. The parameters γ and σ are real numbers, where γ is referred to as the *center of drift* and determines the location. σ represents the Gaussian diffusion part. A Lévy process thus is an arbitrary combination of Gaussian diffusion, a deterministic linear trend and variably designed jump component with finite variation or unbounded variation. These three components are comprised in the so-called *Lévy triplet*.

In case of $\nu(dx) = 0$ this yields the characteristic function of the normal distribution. The scenario that is of interest in this work is the purely non-Gaussian case. It is characterized by setting $\sigma = 0$, yielding as characteristic function of purely non-Gaussian distributions:

$$\exp(i\gamma u + \int\limits_{-\infty}^{+\infty} (e^{iux} - 1 - iux 1_{|x|\leq 1})\nu(dx)).$$

The distributions of interest are the CTS and the NTS distribution are both purely non-Gaussian distributions.

Following Cont and Tankov (2004), there are three basic types of transformation, under which the class of Lévy Processes is invariant, to construct new Lévy Processes. These are linear transformation, subordination and exponential tilting of the Lévy measure. In this master thesis we use subordination to derive the NTS Process. Subordination describes a stochastic time change of an existing stochastic process by a subordinator.

2.3.2. Subordination

Univariate Lévy Processes $(S_t)_{t\geq 0}$ on $\mathbb{R}$ with almost surely non-decreasing trajectories are called *subordinators*. They can be used as time changes for other Lévy Processes. This idea is closely related to the idea of introducing the natural time flow of financial markets to the theory of financial modeling. As financial markets exhibit times of a lot of activity and times of less activity, refered to as business time by several authors. One could say that in the case time is running faster in the other slower. Consequently, a constant timeflow is inappropriate for financial modeling. As time is nondecreasing itself, we also need this property for the subordinator. To express this condition, the following list of equivalent conditions. For Proof we refer to Cont and Tankov (2004).

i) $S_t \geq 0$ a.s. for some $t > 0$.

ii) $S_t \geq 0$ a.s. for every $t > 0$.

iii) Sample paths of (S_t) are almost surely nondecreasing: $t \geq s \Rightarrow S_t \geq S_s$ a.s..

iv) The characteristic triplet of (X_t) has no diffusion component ($\sigma = 0$) and a non-negative additional deterministic linear trend $\mu \geq 0$. Moreover, the Lévy measure ν has to be concentrated on the positive real line and is of finite variation: $\int_0^\infty (x \wedge 1)\nu(\,dx) < \infty$. Only positive jumps of finite variation.

2.3.3. Time-changed Stochastic Process

In case the deterministic time of a stochastic process $Y_{(t)}$ is replaced by a subordinator S_t. Y is said to be subordinated by S. This creates a new stochastic process

$$X_{(t)} = Y_{S_{(t)}}. \tag{2.3.2}$$

In case that $Y_{(t)}$ is a general Lévy Process itself, the outcome $X_{(t)}$ of the subordination again satisfies the condition of a Lévy process.

2.3.4. Time-changed Brownian Motion

For later construction reasons the case of time-changed Brownian Motion is briefly explained in a more general setting. Subordinating Brownian Motion with drift μ and volatility γ we obtain the time-changed Brownian motion. Considering the Arithmetic Brownian Motion[5]

$$\mu t + \gamma W_t$$

Considering a subordinator $(S_t)_{T \geq 0}$ independent to the standard Brownian Motion $(W_t)_{t \geq 0}$. Subordination yields a new Process $X_{(t)}$ with:

$$X_t = \mu S_t + \gamma W_{S_t} \qquad (2.3.3)$$

which is the time changed Brownian Motion. Its characteristic function is given by

$$\phi_{X_t}(u) = \phi_{S_t}\left(\mu u + \frac{iu^2\gamma^2}{2}\right),$$

with ϕ_{S_t} being the characteristic function of S_t. As presented by Clark (1973), every Lévy Process can be viewed as a subordinated Brownian motion. Thus, the extension to the multivariate case is possible. For construction of the NTS process a CTS subordinator is used.

2.3.5. The α-stable Process

Although the normal distribution has been frequently applied in modeling the return distribution of assets, its properties are not consistent with the observed behaviour found for asset returns. The symmetric and rapidly decreasing tail properties of the normal distribution cannot describe the skewed and fat-tailed properties of the empirical distribution of returns. The α-stable distribution has been proposed as an alternative to the normal distribution for modelling asset returns as it allows for the desired properties. As the α-stable distribution represents the basis for the development towards tempered stable distributions, it shall be briefly introduced. We therefore follow Rachev et al. (2011).

Definition 2.2 (Stable Distributions). *A random variable X follows an α-stable distribution if for all $n \in \mathbb{N}$ exist i.i.d. $X_1, ..., X_n$, a positive constant C_n and a real number D_n such that:*

---

[5]We keep the basic theoretical presentation limited to the univariate case but it can be possibly extended to multivariate case with a real vector $\mu \in \mathbb{R}^n$ and a Covariance Matrix Σ . This will be used later for construction of the multivariate NTS model

$$(X_1 + X_2 + ... + X_n) \overset{d}{=} C_n X + D_n.$$

The notation "$\overset{d}{=}$" denotes equality in distribution. The constant $C_n = n^{\frac{1}{\alpha}}$ dictates the stability property. In case of $\alpha = 2$ this yields the Gaussian case. For the general case the density of the α-stable distribution does not have a closed-solution. The Lévy measure of the α-stable distribution is given by

$$\nu_{\text{stable}}(dx) = \left(\frac{C_+}{x^{1+\alpha}} 1_{x>0} + \frac{C_-}{|x|^{1+\alpha}} 1_{x<0} \right) dx. \tag{2.3.4}$$

By using 2.3.1 and 2.3.4 the distribution is expressed by its characteristic function:

$$\phi(u; \alpha, \sigma, \beta, \mu) = E[e^{iuX}]$$

$$\phi_{stable} = \begin{cases} \exp(i\mu u - |\sigma u|^{\alpha}(\text{sgn } u) \tan \frac{\pi\alpha}{2}), & \alpha \neq 1 \\ \exp(i\mu u - \sigma|u|(1 + i\beta\frac{2}{\pi}(\text{sgn } u) \ln |u|)), & \alpha = 1 \end{cases}, \tag{2.3.5}$$

where

$$\text{sgn } t = \begin{cases} 1, & t > 0 \\ 0, & t = 0 \\ -1, & t < 0, \end{cases}$$

which can be found in Rachev and Mittnik (2000). The correspondance of the parameter transformation from $(\alpha, C_+, C_-, \gamma)$ to the one used in 2.3.5 using $(\alpha, \beta, \sigma, \mu)$ is shown in Sato (1999). The distribution is characterized by four parameters:

1. α: $\alpha \in (0, 2)$ the index of stability or the shape parameter,

2. β: $\beta \in [-1, +1]$ the skewness parameter,

3. σ: $\sigma \in (0, +\infty)$ the scale parameter,

4. μ: $\mu \in (-\infty, +\infty)$ the location parameter,

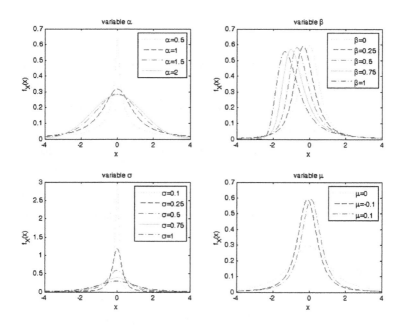

FIGURE 2.2.: *Illustration of the role of different parameters in the α-stable distribution.*

Caused by the four parameters, the α-stable distribution is highly flexible and suitable for modelling non-symmetric, highly kurtoic and heavy-tailed data. For a complete treatment of α-stable distributions and processes it can be referred to Samorodnitsky and Taqqu (2000) and Rachev and Mittnik (2000). Besides these advantages of the α-stable distribution it has some major drawbacks regarding the existence of the moments. This becomes highly relevant in the application process of heavy-tailed distributions for Option Pricing.

Property 2.1. *Raw moments satisfy the property:*

$$E|X|^p < \infty, \qquad\qquad \text{for any } 0 < p < \alpha$$
$$E|X|^p = \infty, \qquad\qquad \text{for any } p \geq \alpha.$$

The fact that α-stable random variables with $\alpha < 2$ have an infinite second moment means that many of the techniques valid for the Gaussian case do not apply. Further Property 2.1 leads to the fact when $\alpha \leq 1$, one also has $E|X| = \infty$, precluding the use of expectations. We refer to Samorodnitsky and Taqqu (2000), section 1.2, for a proof of this result.

To overcome the disadvantage of infinite moments, two solutions have been introduced. One is the smoothly truncated stable distribution by Menn and Rachev (2009), that shall be mentioned for reasons of completeness and not be further discussed. The principle idea can be formulated by cutting the heavy-tails and replacing by thinner tails to achieve finite moments. The other possible solution are the so called tempered stable distributions.

2.3.6. Tempered Stable Distribution

In order to overcome the drawback of infinite moments of all orders, Koponen (1995) introduced the idea of multiplying the α-stable density by an exponentially decreasing so called *tempering function* on each half of the real axis. After this exponential tempering, the small jumps keep their initial stable-like behaviour whereas the large jumps become much less violent. This approach, introduced by Koponen (1995), for the construction of stochastic processes as *truncated Lévy flights*, Boyarchenko and Levendorskiï (2000) as *KoBoL* and finally by Carr et al. (2003) as CGMY and referred to as Classical Tempered Stable (CTS) by Rachev et al. (2011) lauched the development of a broad range of various different types such as the Normal Tempered Stable, the Rapidly Decreasing, the Kim-Rachev Tempered Stable, etc. Due to the common fact of using a tempering function in order to achieve finite moments, they have generally been termed as *Tempered Stable* processes in the literature. Current research is going on for developing a general framework for tempered stable distributions. In order to construct a NTS model subordination can be used. The subordinator is a CTS process. The CTS process has the following Lévy Measure:

$$\nu(x) = \frac{C_+ \exp(-\lambda_+|x|)}{|x|^{\alpha+1}} \mathbb{1}_{>0}(x) + \frac{C_- \exp(-\lambda_-|x|)}{|x|^{\alpha+1}} \mathbb{1}_{<0}(x), \qquad (2.3.6)$$

with parameters

- α: the index of stability or the shape parameter, $\alpha \in (0,1) \cup (1,2)$

- λ_+, λ_-: serve to control the exponential tempering, $\lambda > 0$

Using the Lévy-Khinchine formula (2.3.1), the CTS distribution has the following characteristic function.

$$\phi_X(u) = \phi_{CTS}(u; \alpha, C, \lambda_+, \lambda_-, m)$$

$$= \exp(ium - iuC\Gamma(1-\alpha)(\lambda_+^{\alpha-1} - \lambda_-^{\alpha-1})$$

$$+ C\Gamma(-\alpha)((\lambda_+ - iu)^\alpha - \lambda_+^\alpha + (\lambda_- + iu)^\alpha + \lambda_-^\alpha))$$

2.3.7. CTS Subordinator

In order to fulfill the properties of a subordinator, the Lévy density of the CTS form equation (2.3.6) has to be concentrated on the positive real line first (by setting the negative part to zero with $C_+, \lambda_+ > 0$):

$$\nu(x) = \frac{C_+ \exp(-\lambda_+ x)}{x^{\alpha+1}} \mathbb{1}_{>0}(x). \qquad (2.3.7)$$

. The necessity of finite variation yields $\alpha \in (0, 1)$. α can be reset to its original range by changing the Lévy Measure to

$$\nu(x) = \frac{C_+ \exp(-\lambda_+ x)}{x^{\alpha/2+1}} \mathbb{1}_{>0}(x). \qquad (2.3.8)$$

Inserting (2.3.8) in (2.3.1) with $\gamma = \int_0^1 x\nu(dx)$ the following characteristic function for the CTS Subordinator can be derived:

$$\phi_{S_t}(u) = \exp\left(tC_+ \int\limits_0^\infty (e^{iux} - 1)\frac{e^{-\lambda_+ x}}{x^{\alpha/2+1}} dx \right).$$

Solving the integration

$$\phi_{S_t}(u) = \exp\left(tC_+ \Gamma\left(-\frac{\alpha}{2}\right) \left((\lambda_+ - iu)^{\frac{\alpha}{2}} - \lambda_+^{\frac{\alpha}{2}}\right) \right)$$

is derived. For easier handling and due to the fact of the non existing negative part of the Lévy measure we set $C_+ = C$ and $\lambda_+ = \theta$ in the rest of the master thesis. Thus, the characteristic function of the subordinator is

$$\phi_{S_t}(u) = \exp\left(tC\Gamma\left(-\frac{\alpha}{2}\right) \left((\theta - iu)^{\frac{\alpha}{2}} - \theta^{\frac{\alpha}{2}}\right) \right). \qquad (2.3.9)$$

In order to avoid construction redundancies, meaning that the subordinator can be constructed in a variety of ways by varying $C = \frac{\theta^{1-\alpha/2}}{\Gamma(1-\alpha/2)}$. Inserting in (2.3.9) we receive

$$\phi_{S_t}(u) = \exp\left(-t\frac{2\theta^{1-\alpha/2}}{\alpha} \left((\theta - iu)^{\alpha/2} - \theta^{\alpha/2}\right) \right). \qquad (2.3.10)$$

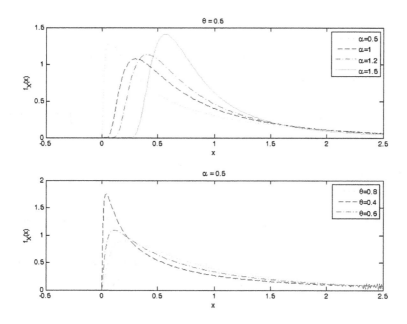

FIGURE 2.3.: *CTS Subordinator - Influence of θ, α*

2.3.8. Univariate Normal Tempered Stable Distribution

The univariate *Normal Tempered Stable (NTS)* process $X_{(t)}$ in the univariate case is defined by replacing the physical time t of the univariate Brownian Motion by a stochastic time flow using the CTS Subordinator from Section 2.3.7.

As focus is first only on single distributions. The stochastic time index $S_{(t)}$ can be seperated from the Brownian Motion[6]. Thus $X \sim NTS(\alpha, \theta, \beta, \gamma, \mu)$ can be defined by

$$X = \mu + \beta(S - 1) + \gamma\sqrt{S}W, \tag{2.3.11}$$

where $\mu, \beta \in \mathbb{R}, \gamma > 0, W \sim N(0,1)$, and S the CTS subordinator with parameter (α, θ). The characteristic function of X is given by

$$\phi_X(u) = \exp\left(i(\mu - \beta)u - \frac{2\theta^{1-\alpha/2}}{\alpha}\left((\theta - i(\beta u + \frac{i\gamma^2 u^2}{\alpha}))\right)^{\alpha/2} - \theta^{\alpha/2}\right). \tag{2.3.12}$$

2.3.9. Standard Univariate NTS

In the context of modelling the situation often occurs where the considered data is assumed to have zero mean and unit variance. Standardization can be achieved via linear transformation, mass scaling or parameter manipulation. For standardization we demonstrate the procedure of parameter manipulation. The first step is to set $E(X) = 0$ by choosing $\mu = 0$. In order to achieve unit variance we set

$$\gamma = \sqrt{1 - \beta^2\left(\frac{2 - \alpha}{2\theta}\right)}.$$

In order to achieve a feasible Brownian motion, $\gamma > 0$ and

$$|\beta| < \sqrt{\left(\frac{2\theta}{2 - \alpha}\right)}$$

has to hold. The specific values of parameters can be represented as functions of the remaining free distribution parameters (α, θ, β):

$$\mu(\alpha, \theta, \beta) = 0,$$

$$\gamma(\alpha, \theta, \beta) = \sqrt{1 - \beta^2\left(\frac{2 - \alpha}{2\theta}\right)}.$$

---

[6]For an extensive derivation of the connection between (2.3.3) and (2.3.11), we refer to the dissertation of Krause (2011).

Inserting this in the definition of the NTS distribution yields the definition of the *standard Normal Tempered Stable distribution (stdNTS)*

$$stdNTS(\alpha, \theta, \beta) \equiv NTS\left(\alpha, \theta, \beta, 1 - \beta^2 \left(\frac{2 - \alpha}{2\theta}\right), 0\right).$$

with characteristic function

$$\phi_X(u) = \exp\left(-\beta u i - \frac{2\theta^{1-\alpha/2}}{\alpha}\left(\left(\theta - i\beta u + \left(1 - \beta^2 \left(\frac{2 - \alpha}{2\theta}\right)\right)\frac{u^2}{2}\right)^{\alpha/2} - \theta^{\alpha/2}\right)\right) \quad (2.3.13)$$

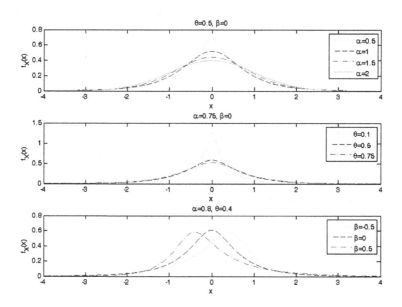

FIGURE 2.4.: *Illustration of the role of different parameters in the NTS distribution*

2.3.10. Multivariate Standard Normal Tempered Stable Distribution

For the construction of a multivariate Standard Normal Tempered Stable distribution, the same approach as for the univariate Normal Tempered Stable distribution can be applied. The reason for limiting the presentation to the Multivariate Standard Normal Tempered Stable case is due to the fact that the standardized case is relevant in connection with an ARMA-GARCH time series model as presented in section 3.4.3.

The CTS Subordinator from section 2.3.7 serves as a stochastic time change to each single component of a now multivariate Brownian motion with a possible dependence between the dimensions. With $S_{(t)}$ and $B_{(t)}$ being independent and given $\mathbf{X} \in \mathbb{R}^n$

$$\mathbf{X}_{(\mathbf{t})} = (X_{(t)}^1, X_{(t)}^2, ..., X_{(t)}^n)^T. \tag{2.3.14}$$

Assuming $\mu = (\mu_1, ..., \mu_n) \in \mathbb{R}^n, \beta = (\beta_1, ..., \beta_n) \in \mathbb{R}^d$ and $\Sigma \in \mathbb{R}^{d \times d}$ being a positive definite matrix and $B_{(t)}$ has covariance matrix Σ. The distribution is defined as the distribution of the process variable after unit time ($t = 1$). Further, we say that X_1 has MNTS distribution with parameters $(\alpha, \theta, \beta, \gamma, \mu, \rho)$ if

$$X = \mu + \beta(S - 1) + diag(\gamma)\sqrt{S}W, \tag{2.3.15}$$

with $W \sim N(0, I)$ denoting a d-dimensional standard normal distribution.[7] In order to meet the required conditions of a standardized distribution we put $\mu = (0, ..., 0)^T$ which centers the distribution around zero and in order to achieve unit variance we set for all $k \in \{1, 2, ..., n\}$

$$\gamma_k = \sqrt{1 - \frac{2 - \alpha}{2\theta}\beta_k^2},$$

with

$$|\beta_k| < \sqrt{\frac{2\theta}{2 - \alpha}}.$$

---

[7]For the derivation of (2.3.15) we refer to Krause (2011).

Then

$$E(X) = (0, ..., 0)^T \tag{2.3.16}$$

$$Var(X) = (1, ..., 1)^T \tag{2.3.17}$$

$$Cov(X) = (Cov(X^i, X^j))_{i,j \in \{1,2,...,n\}} \tag{2.3.18}$$

$$= \operatorname{diag}(\gamma)\rho\operatorname{diag}(\gamma) + \frac{2-\alpha}{2\theta}\beta\beta^T. \tag{2.3.19}$$

This distribution is called *Mulitvariate Normal Tempered Stable Distribution* and we denote it by stdMNTS($\alpha, \theta, \beta, \rho$).

The characteristic function of the MNTS distribution is given by

$$\phi_X(u) = \exp\left(i(\mu - \beta)^T u - \frac{2\theta^{1-\alpha/2}}{\alpha}\left(\left(\theta - i\beta^T u + \frac{1}{2}u^T \Sigma u\right)^{\alpha/2} - \theta^{\alpha/2}\right)\right),$$

$$\forall u \in \mathbb{R}^n,$$

yielding for the standardized case

$$\phi_X(u) = \exp\left(-i\beta^T u - \frac{2\theta^{1-\alpha/2}}{\alpha}\left(\left(\theta - i\beta^T u + \frac{1}{2}u^T \Sigma u\right)^{\alpha/2} - \theta^{\alpha/2}\right)\right),$$

$$\forall u \in \mathbb{R}^n.$$

$$\gamma = \left[\sqrt{1 - \frac{2-\alpha}{2\theta}\beta_k^2}\right]_{k=1,...,n}^T.$$

2.4. Goodness of Fit

After a model has been chosen and applied to the data (i.e., the models parameters have been estimated), it is crucial to verify the model's descriptive power. This includes especially a goodness-of-fit examination for the employed innovation distribution assumption. To this end, several statistical tools are available. The proposed goodness of fit test will result in a p-value which has to be compared to predefined confidence level, typically 95%, 99% or 99.9%. The hypothesis that the considered model is consistent with the observed data us usually denoted the *Null Hypothesis* (H_0). If the resulting p-value falls below the predefined level, the Null Hypothesis is rejected, otherwise it is accepted. A general issue which arises when using statistical tests are two specific error types.

Type I Error: Rejecting the Null-Hypothesis when it is actually true.

Type II Error: Accepting the Null Hypothesis when it is actually false.

The choice of the confidence level is a mean to find a suitbale compromise between Type I and Type II errors. It is not possible to rule out both errors at the same time. Being conservative in the acceptance of a model will reduce the probability of Type II Errors but at the same time increase the probability of Type I Errors and vice versa. The 95% confidence level is rather conservative, thus reduces the probability of Type II Errors whereas the 99.9% confidence level reduces the probability of Type I Errors.

The goodness-of-fit tests which have been used in this thesis are presented in this section.

2.4.1. Kolmogorov-Smirnov-Test

The Kolmogorov-Smirnov (KS) test is a nonparametric test that assess if two cumulative probability density functions (cdf) result from identical distributions within statistically allowed boundaries. Given two probability distributions F and $\hat{F}$. The Null Hypothesis is defined by

$$H_0 : F(x) = \hat{F}(x),$$

We cann assess a distance between these two distrubutins by calculating the highest distance between the values $F(x)$ and $\hat{F}(x)$ for varying x. Mathematically thsi means calculating the supremum distance between $F(x)$ and $\hat{F}(x)$. The distance is used as a test statistic and is constructed by:

$$KS = \sqrt{n} \sup_{x_i} |F(x_i) - \hat{F}(x_i)|, \qquad (2.4.1)$$

where $F(x)$ is the theoretical cumulated distribution function (cdf) and $\hat{F}(x)$ denotes the empirical sample distribution. The KS- statistic turns out to emphasize deviations around the median of the fitted distribution.

2.4.2. Anderson-Darling Test

The Anderson-Darling (AD) test is carried out by following a similar procedure as compared to the KS test but with a modified test statistic. The AD statistic is computed by

$$AD = \sup_{x \in \mathbb{R}} \frac{|F(x_i) - \hat{F}(x_i)|}{\sqrt{\hat{F}(x_i)(1 - \hat{F}(x_i))}} \qquad (2.4.2)$$

The AD-statistics scales the absolute deviations between empirical and fitted distributions with the standard deviation of $F(x)$. By its construction, the AD-statistic accentuates more the decrepancies in the tails.

2.5. Risk Management

There is no universial definition of what constitutes risk, although risk management plays an utmost important role for both individuals and corporate entities. On a macro perspective risk could be defined as the chance of losing part or all of an investment. Due to the fact that financial and non-financial instutions own complex portfolios composed of many financial contracts and given the reoccuring financial disasters and times of distress brought to the fore the necessity and importance of efficient methods for measuring and controlling risk. Because financial risks often manifest themselves in subtle and nonlinear ways in corporate balance sheets and income statements the focus is on quantifying risk in a statistical sense (Aït-Sahalia and Lo 2000). While there are many sources of financial risk, we concentrate here on *market risk*, meaning the risk of unexpected changes in prices or rates. Other forms of financial risk include liquidity risk, credit risk, operational risk and further types of risk. In this section we introduce two risk measures, the industry standard risk measure, *Value-at-Risk*, and the *Average Value-at-Risk*.

2.5.1. Value-at-Risk

A standard benchmark risk measure that has been widely accepted since the 1990s is the Value-at-risk (VaR)[8]. In the late 1980s, it was used by major financial firms to measure the risks of their trading portfolios. J.P. Morgan established a market standard through its RiskMetrics system in 1994. In the mid 1990s the VaR measure was appoved by regulators as a valid approach to calculating capital reserves needed to cover market risk. The Basel Committee on Banking Supervision released a package of amendments to the requirements for banking institutions allowing them to use their own internal systems of risk estimation. In this way, capital reserves, which financial institutions are required to keep, could be based on the VaR numbers computed internally by an in-house risk management system. Generally, regulators demand that the capital reserves equal the VaR number multiplied by 3 and 4. Thus, regulators link the capital reserves for market risk directly to the risk measure. There are three key elements of VaR - specified level of loss in value, a fixed time period over which risk is assessed and a confidence interval. The VaR can be specified for an individual asset, a portfolio assets or for an entire firm. The

---

[8]For a general overview of Value at Risk we refer to Duffie and Pan (1997) and Linsmeier and Pearson (2000)

recommended confidence levels are 95% and 99%. Suppose that we hold a portfolio with a one-day 99% VaR equal to $1 million. This means that over the horizon of one day the portfolio may lose more than $ 1 million with probability of equal to 1%. The same example can be constructed for percentage returns. VaR is a universal concept and can be applied to most financial instruments. It is a distribution independent single summary statistical measure of risk. Denote by $(1 - \epsilon)100\%$ the condfidence level parameter of the VaR. As we explained, losses larger than the VaR occur with probability ϵ. The probability ϵ is called tail probabiliy. What sometimes leads to confusion is the different existing definitions of the VaR and the number resulting from the different definitions. In this thesis we define the VaR in the following way. VaR at confidence level $(1 - \epsilon)100\%$ is defined as the negative of the lower ϵ-quantile of the return distribution,

$$VaR_\epsilon(X) = -\inf x | P(x \leq x) \geq \epsilon = -F_X^{-1}, (\epsilon) \tag{2.5.1}$$

where $\epsilon \in (0, 1)$ and $F^{-1}(\epsilon)$ is the inverse of the distribution function. If the random variable X describes random returns, then the VaR number is given in terms of a return figure. Although VaR is a very intuitve concept its measurement is a challenging statistical problem. Computation approaches for the VaR can be classified into two groups: parametric approaches and non-parametric approaches. All methods have their own strengths and weaknesses and the problem that they might yield different results. The non-parametric VaR is obtained from historical data and the two parametric approaches are the Delta-normal method or variance-covariance method and the Monte Carlo simulations method (Hull and White 1998)[9].

While Value at Risk has aquired a strong following in the risk management community, there is reason to be skeptical of both its accuracy as a risk management tool and its use in decision making. Danielsson (2011) states three main issues that arise in the implementation of VaR:

1. VaR is only one quantile.

2. VaR is not a coherent risk measure.[10]

3. VaR is easy to manipulate.

---

[9]2.5.4

[10]2.5.2

2.5.2. Coherent Risk Measures

Artzner et al. (1998) study the properties a risk measure should have in order to be considered a sensible and useful risk measure. In their paper they present and justify a set of four desirable properties for measures of risk, and call the measures satisfying these properties *coherent*. Let a risk measure be denoted by $\varphi(\cdot)$. Consider two real-valued random variables X, Y. A function $\varphi(\cdot) : X, Y \rightarrow \mathbb{R}$ is called a coherent risk measure if it satisfies for X, Y and a constant c the following properties:

1. Translation invariance: For all $c \in \mathbb{R}$, $\varphi(X + c) = \varphi(X) - c$.

   Adding c to the portfolio is like adding cash, which acts as insurance, so the risk of $X + c$ is less than the risk of X by the amount of c

2. Subadditivity: $\varphi(X + Y) \leq \varphi(X) + \varphi(Y)$.

   The risk of the portfolios of X and Y cannot be worse than the sum of the two individual risks.

3. Positive homogeneity: For all $c \geq 0$, $\varphi(cX) = c\varphi(X)$.

   By changing all positions in a portfolio by the factor c while keeping the relative proportions constant, the risk measure must change by the same factor c.

4. Monotonicity: If $Y \geq X$, then $\varphi(Y) \leq \varphi(X)$.

   If portfolio X never exceeds the values of portfolio Y, the risk of Y should never exceed the risk of X.

As stated in Section 2.5.1 VaR is not a coherent risk measure, since it does not always satisfy the axiom of subadditivity.[11] VaR is, however, subadditive in the special case of normally distributed returns. Danielsson et al. (2010) study the subadditivity of VaR further and find that VaR is indeed subadditive provided the tail index exceeds 2. For sufficiently fat tails subadditivity does not hold for VaR.

2.5.3. Average Value-at-Risk

An additional measure of risk closley related to VaR is the average Value-at-Risk (AVaR), also called Expected Shortfall (ES) or Conditional VaR (CVaR) or Expected Tail Loss (ETL). AVaR

---

[11]This can be demonstrated on behalf of examples. See (Danielsson 2011),(Hull 2007).

is a risk measure that is a superior alternative to VaR. Not only does it lack the deficiencies of VaR, but it also has an intuitive interpretation. There are convenient ways for computing and estimating AVaR that allows its application in optimal portfolio selection problems. Moreover, it satisfies all axioms of coherent risk measures and is consistent with the preference relations of risk-averse investors. Furthermore, not only does AVaR provide information about losses beyond VaR, it is a convex, smooth function of portfolio weights and is therefore attractive as a risk measure for optimizing portfolios. The AVaR at tail probability δ is defined as the average of the VaRs that are larger than the VaR at tail probability δ. Therefore, by construction, AVaR is focused on the losses in the tail that are larger than the corresponding VaR level.

Let X be a real random variable in L^1, that is $E[X] < \infty$. Then we define the risk measure AVaR as the following convergent integral

$$AVaR_\delta(X) := \frac{1}{\delta} \int\limits_0^\delta VaR_p(X) \, \mathrm{d}p \qquad (2.5.2)$$

Unfortunately, due to the definition of AVaR, a closed-form formula for the valuation of equation 2.5.2 is only available for a few distributions.

2.5.4. Computation

For computation of risk measures there exist, as mentioned, three different approaches. The historical method, the variance-covariance method and the Monte Carlo approach.

The historical method simply consists of using the percentiles of the actual historical returns to calculate the VaR. All anomalies of past events such as fat-tails, skewness and kurtosis are included. Each historical observation carries the same weight in the historical method. It implies that the future is assumed to behave like the past, in other words assuming stationarity of the loss distribution. This is not verified in reality, especially leading to wrong results in the case of structural breaks.

The variance-covariance method assumes a probability distribution. Historical data is used for calibration of the distributional parameters. Therefore, a closed-form formula of the risk measure is necessary. It is a linear approach, as it is assumed that instrument prices change linearly with respect to changes in the risk factors. It is therefore not capable of capturing nonlinear effects, such as nonlinear payoffs or calls. The computation is fast and can capture events not present in historical data.

The Monte Carlo Method is intended to simulate the future evolution of risk factors. A probability distribution is defined for each risk factor and parameters of each distribution are estimated on the basis of the past of these risk factors. Then a large number of outcomes is simulated. This method is suitable for all types of underlying portfolios but very computationally intensive. Therefore it tends to be unsuitable for large, complex portfolios.

2.5.5. Backtesting

Backtesting is a procedure that can be used to compare the various risk models and also formes the basis of the current regulatory framework. Backtesting is useful in identifying the weakness of risk-forecasting models in comparing an ex ante risk forecast from a particular model and comparing them with ex post realized returns.

Since the late 1990's a variety of tests have been proposed to test for accuracy of a model.[12] While many of these tests differ in their details they focus on the situation whenever losses exceed the risk measure, a violation is said to have occured. Denoting VaR as the chosen risk measure and the profit of loss on the portfolio over a fixed time interval as $x_{t,t+1}$. The "hit" function can be defined as

$$I_{t+1}(\epsilon) = \begin{cases} 1 \ if \quad x_{t,t+1} \leq -VaR_t(\epsilon) \\ 0 \ if \quad x_{t,t+1} > -VaR_t(\epsilon). \end{cases} \qquad (2.5.3)$$

In order to have an accurate VaR model, the hit sequence must satisfy the following two properties (Christoffersen 1998):

1. Unconditional Coverage Property - The probability of realizing a loss in excess of the reported VaR, $VaR_t(\epsilon)$ must be precisely $\epsilon \times 100\%$. In cases that losses excess the reported VaR measure more frequently would suggest a systematical underestimation of the actual level of risk. The opposite would alternatively suggest a too conservative risk measure.

---

[12]For a good overview on backtesting of VaR, see (Lopez 1999) and (Campbell 2005).

2. Independence Property - Any two elements of the hit sequence, $I_{t+j}(\epsilon), I_{t+k}(\epsilon)$ must be independent from each other. Thus, a clustering of VaR violations represents a violation of the independence property that signals a lack of responsiveness in the reported VaR measure.

Some of the earliest proposed VaR backtests, e.g. Kupiec (1995), focused exclusivly on the property of unconditional coverage. In the light of the failure of tests of unconditional coverage to detect violations of the independence property, a variety of tests have been developed like Christoffersen (1998) or Berkowitz (2001). [13]

---

[13]for futher details we refer to the specific papers Christoffersen (1998) or Berkowitz (2001)

3. Data and Methodology

"In God we trust; all others must bring data"

W. Edwards Deming

This chapter gives detailed insights into the methodological approach and used data for all upcoming analysis. Section 3.1 presents characterisitics for the data set. Section 3.2 describes transformation of the data which is applied to remove the intraday volatility profile. Subsequently, ARMA-GARCH and FIGARCH models with different innovation distribution assumptions are compared in their modeling capabilities. The special case of tempered infinitely divisible innovation distributions is investigated further to find dependencies between the corresponding parameter values and the frequency on which log-returns are constructed. The forecasting performance of the respective models is investigated in a Value-at-Risk backtest.

The empirical analysis was carried out by employing MATLAB and SAS.

3.1. Data Selection

Data used in this thesis is obtained from Thomson Reuters DataScope Tick History (TRDTH) [1] archive through the Industry Research Centre of Asia Pacific(SIRCA).[2] TRDTH holds trade and quote data (TAQ) for "more than 45 million unique instruments across 400+ exchanges" timestamped up to milliseconds.[3] In this thesis I use aggregated data as already provided by TRDTH. The data is aggregated for fixed intraday periods, e.g. 1 min. In particular, retrieved data includes Reuters Identification Code (RIC), opening price, highest price, lowest price, last price, volume for the aggregation period.

The thesis is based on the most important German stock index DAX which consists of the 30 largest companies listed on Frankfurt Stock Exchange. To be regarded as a potential candidate for being taken into the DAX, companies need to fulfill a number of specified requirements. Likewise, companies can be removed from the DAX if they violate certain criteria. The DAX value is constructed by applying a weighted sum of each company's market capitalization. The weight used in the index' construction is proportional to the number of publicly available shares. To have the index' time series free of distortions resulting from company removals, rights issues, issuances etc., the index' value is re-adjusted after such events.

The DAX belongs to the group of total return indices. That is, dividend payments are reinvested.

For the DAX we have a minute-by-minute set from 2003 to 2012 where the trading day is from 8:00 to 16:45 (UTC[4]). Weekends and major holidays were deleted for all data sets an in case of intraday series the first observation of each day as well, in order to avoid overnight effects. Our database covers almost all entire business days of the year except for rare failures. To the best of our knowledge, the data is checked for erroneous observations. Missing values are obtained by interpolation. Whereas the data quality provided by Reuters is already very high.

---

[1]http://thomsonreuters.co/products_services/financial/financial_products/a-z/tick_history/.
[2]I thank the IISM of the Karlsruhe Institute of Technology for providing access to Thomson Reuters DataScope Tick History archive im.uni-karlsruhe.de.
[3]Citation taken from http://thomsonreuters.co/products_services/financial/financial_products/a-z/tick_history/.
[4]Short for 'Universal Time Coordinated' which is one hour delayed to Central European Time.

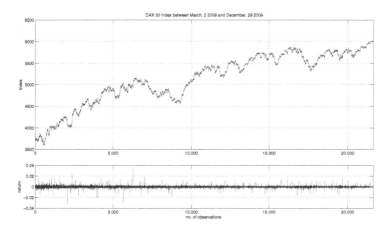

FIGURE 3.1.: *Illustration of the DAX 30 Index during the analyzed time period between March,*
2 2009 and December, 29 2009

3.2. Data Transformation

A direct application of traditional time series methods to raw high frequency returns may give

rise to erroneous inference about the return volatility dynamics (Andersen and Bollerslev 1997).

It is widely documented that return volatility varies systematically over the trading day and that

this pattern is highly correlated with intraday variation of bid-ask spreads and trading volume

(Fraenkle 2010). The empirical evidence on the properties of average intraday stock returns dates

back to Wood et al. (1985) and (Harris 1986) who document the existence of a distinct U-shape

pattern (Figure 3.2) in return volatility over the trading day. That is, volatility is generally

higher in the openening and closing of trading compared to the middle of the day. If this pattern

is ignored, it could potentially misspecifiy an underlying GARCH process. Thus, before ARMA-

GARCH models can be applied to high frequency data, adjusting for the pronounced periodic

structure is critical. Moreover, it is of fundamental importance in uncovering the complex

link between the short- and long-run return components, which may help to explain the conflict

between the long-memory volatility characteristics observed in intraday data and the rapid short-run decay associated with news arrivals in intraday data (Andersen and Bollerslev 1997). Figure

3.3 for the DAX 30 absolute log-returns is equally telling, as it shows how the strong intraday

pattern induces a distorted U-shape in the sample autocorrelation[5] The U-shape occupies exactly

100 intervals, corresponding to the daily frequency.

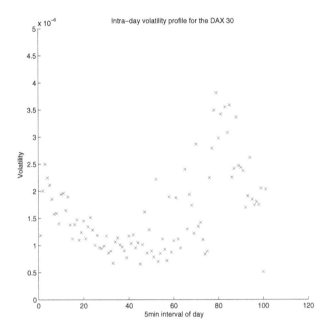

FIGURE 3.2.: *Intra-daily volatility profile for the DAX 30 index on 5 min log-returns for the analyzed time period between March, 2 2009 and December, 29 2009. The sum of squared returns serve as a proxy for the unobserved volatility in this chart.*

To follow in this direction, we apply the method proposed by Beck (2011), which represents a

modified approach from the one Bollerslev et al. (2006) employ in their study. The volatility

---

[5]A corresponding figure is presented in (Andersen and Bollerslev 1997).

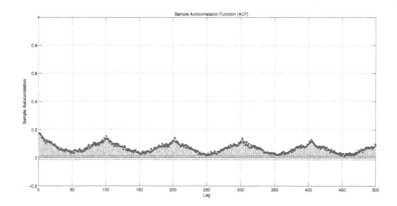

FIGURE 3.3.: *Sample Autocorrelation Function of the absolute log-returns for the analyzed time period between March, 2 2009 and December, 29 2009.*

$v_{d_0,i}$ for return r_i on day d_0 is estimated by the following expression:

$$v_{d_0,i}^2 = \frac{\sum\limits_{k=1}^{N} r_{i,d_{-k}}^2 \cdot \exp(-\tau \cdot k)}{\sum\limits_{k=1}^{N} \exp(-\tau \cdot k)}, \qquad (3.2.1)$$

where $r_{i,d_{-k}}^2$ is a proxy for the unobserved volatility[6] belonging to the ith return on the kth day before day d_0. Due to this modification, the resulting volatility measure will depend more heavily on those previous days, which are close to the day under investigation. By this, the volatility measure adjusts more quickly to persistent changes in the volatility profile's structure and magnitude. In this exponential smoothing model, the parameters are set equivilantly to Beck (2011), $\tau = 0.1$ and $N = 10$. $N = 10$ is chosen such that a time horizon of two business weeks is used to forecast the volatility and $\tau = 0.1$ has been chosen such that the 10th day still has a non-vanishing influence on the volatility. This has as consequence that the weights on the past days are such that days further in the past are weighted less than the days that are closer

---

[6]Volatility itself cannot be observed. A common proxy for the unobserved volatility is the squared return or absolute return.

to the day under investigation.

Following, all returns are transformed by $\hat{r}_{d,i} = \frac{r_{d,i}}{v_{d,i}}$, representing a standardization of intraday

returns, where $\hat{r}_{d,i}$ denotes the transformed return and $r_{d,i}$ denotes the ith observed return on

day d. The result is a time series free of an intraday volatility profile.

3.3. Autocorrelation and Dependence

The sample autocorrelation function (SACF), the sample partial autocorrelation function (SPACF)

and the sample autocorrelation function of log-returns of DAX 30 high-frequency data are dis-

cussed in this section. The examination of these functions allows to analyze autocorrelation

and dependency patterns in the time series and to demonstrate the influence of the exponential

smoothing.

Figure 3.4 illustrates the SACF and SPACF for log-returns as well as the SACF for absolute

and squared log-returns for DAX 30 time series for 5 min high frequency data for the analyzed

time period between March, 2 2009 and December, 29 2009. For the log-retruns despite being

significant, their values remain quite low, thus only indicating weak autocorrelation effects. Long

range dependency effects do not seem to be existing since none of the SACFs constructed from

log-return time series exhibit persistent behavior. The SACFs of absolute and squared returns

show persistent instead of exponentially decreasing behavior. This leads to the conclusion that

the squared log-returns, which are proxies for the unobserved volatility, exhibit long range de-

pendency effects. The same conclusion can be drawn from the R/S Statistics shown in figure

3.5, where absolute returns are used as a proxy for the unobserved volatility. This results in a

motivation to apply FIGARCH models in the process of time series analysis.

This motivates the employment of FIGARCH models which are capable of capturing those

effects. The U-shape existing in the raw data, representing a seasonality effect, is due to expo-

nential smoothing no longer existent.

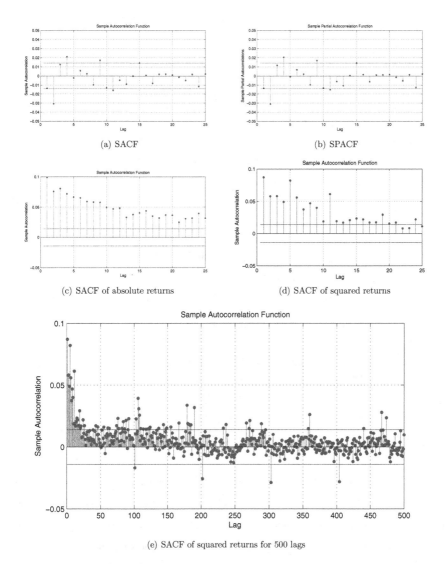

(a) SACF

(b) SPACF

(c) SACF of absolute returns

(d) SACF of squared returns

(e) SACF of squared returns for 500 lags

FIGURE 3.4.: *SACF and SPACF for log-returns as well as the SACF for absolute and squared log-returns for DAX 30 time series for 5 min high frequency data for the analyzed time period between March, 2 2009 and December, 29 2009. They exhibit significant autocorrelation as well as significant and persistent dependency effects.*

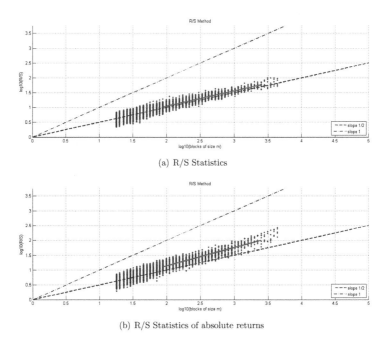

(a) R/S Statistics

(b) R/S Statistics of absolute returns

FIGURE 3.5.: *R/S Statistics for log-returns as well as for absolute log-returns for DAX 30 time series for 5 min high frequency data for the analyzed time period between March, 2 2009 and December, 29 2009.*

3.4. Empirical Estimation and GoF Tests

This part of the thesis is intendend to provide a field application of the multivariate Normal Tempered Stable distribution within the framework of financial data modeling. This includes parameter estimation as well as numerical approximations of probability density (PDF) and cumulated density functions (CDF). We apply it to different time series models based on the data, described in Section (3.1). The goodness of fit of the estimated models will be assessed by a range of powerful statistical standard tests. The section can be split in two parts the first covers the application on index returns, thus in the univariate setting. The second part is focused on the multivariate setting of the DAX 30 index.

3.4.1. Index Returns

In this section, three different times series models (as defined in Section 2.1) are compared on their performance on the DAX index log return series between March, 2 2009 and December, 29 2009 for 5 min return data. We compare the ARMA(1,1)-GARCH(1,1) model with Normal innovation distributions and NTS innovation distribution and the ARMA(1,1)-FIGARCH(1,d,1) with NTS innovation. As shown in Section (3.2) the intraday U-shape of high-frequency data, has to be taken into account and the index log return series is corrected for this effect. Based on experience the assumption of Normal innovations does not hold, even on a daily level. Followingly it can be assumed that for intraday high-frequency data the statistical evidence for rejecting these hypothesis will be even stronger. In order to estimate the parameters of the distribution we use a maximum likelihood(ML) approach. The (quasi-) maximum-likelihood method is used to estimate the model parameters. The NTS-ARMA-(FI)GARCH approach is carried out in a way as described in Kim et al. (2011): (1) estimate ARMA and GARCH parameters using the t-ARMA-GARCH model; (2) extract innovations, and; (3) fit the standard NTS parameters using the extracted innovations. For assessing the Goodness of Fit (GoF) of the estimated distribution the Kolomogorov-Smirnov test (KS), the Anderson-Darling test (AD), as well as the quadratic Anderson-Darling test (AD^2) are used. While the KS test pays more attention to the center of the distribution, both Anderson-Darling tests are designed for detecting deviations in the tails. Starting with a visual examination first, we compare the standardized empirical index return's kernel density, the probability density of the fitted stdNTS with parameters $\alpha = 0.7170, \theta = 0.2965$ and $\beta = -0.0743$ and the standard Normal distribution.

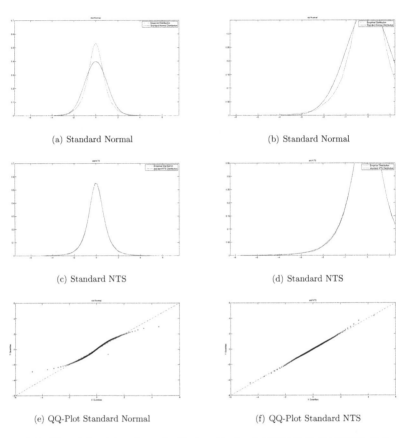

(a) Standard Normal   (b) Standard Normal

(c) Standard NTS   (d) Standard NTS

(e) QQ-Plot Standard Normal   (f) QQ-Plot Standard NTS

FIGURE 3.6.: *Kernel densities of standardized residuals of ARMA(1,1)-GARCH(1,1) in comparison with the fitted standard Normal Distribution and the fitted standard NTS Distribution and QQ-Plots for DAX 30 time series for 5 min high frequency data for the analyzed time period between March, 2 2009 and December, 29 2009.*

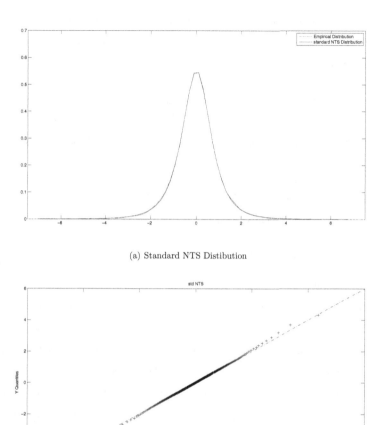

(a) Standard NTS Distibution

(b) Standard NTS QQ-Plot

FIGURE 3.7.: *Kernel densities of standardized residuals of ARMA(1,1)-FIGARCH(1,0.24, 1), compared to fitted standard NTS Distribution and QQ-Plots for DAX 30 time series for 5 min high frequency data for the analyzed time period between March, 2 2009 and December, 29 2009.*

Figure 3.6 shows how the stdNTS follows the empirical return distribution in a more appropriate way compared to the standard Normal distribution. Figure 3.7 further demonstrates the better tail fit of the stdNTS distribution. The two fitted distributions generate the the test statistics presented in table (3.1).

	KS (p-value)	AD	AD^2
ARMA-GARCH Normal	0.0575 (0.0000)	∞	∞
ARMA-FIGARCH NTS	0.0142 (0.1026)	0.0381	2.2320E-4

TABLE 3.1.: *Results of Goodness-of-fit tests for empirical index returns*

The normal-ARMA-GARCH model clearly fails to describe the underlying log return time series, whereas the models with NTS innovation show a better acceptance rate.

3.4.2. Stock Returns

The promising results obtained for the index returns in favour of normal tempered stable distributions should now be further substantiated. The estimation of the different models is performed on the 30 DAX stocks, as members of the DAX between 23.03.2009 and 21.09.2009 as shown in Appendix A. Before and after the composition of the DAX had been and is different. For the 30 DAX stocks we use 5 min high frequency data for the analyzed time period between April, 1 2009 and September, 1 2009. Basic statistics for each of the 30 return series are shown in 3.2. In order to test for the long range dependency we apply the R/S-statistics on the logreturns as well as the absolute returns. In order to take into account the intraday U-shape we alike apply the method proposed by Beck (2011). Table 3.4.2 gives an overview of the 30 DAX stocks and some descriptive statistics for the (log-)returns.

TABLE 3.2.: Descriptive Statistics DAX 30 stocks

	mean	p-value(mean)	confidence interval	std dev	kurtosis	$H_{logreturn}$	$H_{absreturn}$
ADSG	0.000024	0.154477	[-0.000009, 0.000056]	0.002452	193.925213	0.507092	0.507340
ALVG	0.000025	0.178729	[-0.000011, 0.000061]	0.002718	40.225985	0.511780	0.511597
BASF	0.000033	0.055332	[-0.000001, 0.000066]	0.002519	69.027380	0.521375	0.521542
BAYG	0.000033	0.055332	[-0.000001, 0.000066]	0.002519	69.027380	0.521375	0.521542
BEIG	0.000016	0.190988	[-0.000008, 0.000039]	0.001779	80.405454	0.535299	0.534575
BMWG	0.000023	0.214848	[-0.000013, 0.000060]	0.002764	40.656432	0.539724	0.539664
CBKG	0.000036	0.234582	[-0.000024, 0.000097]	0.004522	69.865633	0.538066	0.537624
DAIGn	0.000034	0.104458	[-0.000007, 0.000076]	0.003131	42.722926	0.514510	0.514209
DB1Gn	0.000022	0.254198	[-0.000016, 0.000061]	0.002893	57.747392	0.529662	0.529977
DBKG	0.000042	0.058903	[-0.000002, 0.000086]	0.003282	48.597288	0.537476	0.537189
DPWGn	0.000042	0.058903	[-0.000002, 0.000086]	0.003282	48.597288	0.537476	0.537189
DTEGn	0.000004	0.747497	[-0.000019, 0.000027]	0.001723	216.317363	0.523570	0.523610
EONGn	0.000018	0.248225	[-0.000012, 0.000048]	0.002286	137.046571	0.511199	0.510992
FMEG	0.000006	0.576856	[-0.000015, 0.000027]	0.001567	32.654214	0.525254	0.525370
FREG	0.000008	0.597328	[-0.000023, 0.000040]	0.002346	35.228090	0.512268	0.512851
HNKGp	0.000032	0.020971	[0.000005, 0.000058]	0.002018	351.113362	0.496642	0.496613
HNRGn	0.000008	0.682142	[-0.000031, 0.000047]	0.002908	34.680596	0.522743	0.523286

TABLE 3.2.: *Descriptive Statistics DAX 30 stocks*

	mean	p-value(mean)	confidence interval	std dev	kurtosis	$H_{logreturn}$	$H_{absreturn}$
LHAG	0.000015	0.369568	[−0.000017, 0.000047]	0.002414	662.712308	0.550907	0.552000
LING	0.000024	0.082086	[−0.000003, 0.000051]	0.002032	41.093399	0.497675	0.497837
MANG	0.000026	0.204285	[−0.000014, 0.000066]	0.003020	47.808060	0.518790	0.518905
MEOG	0.000029	0.044335	[0.000001, 0.000058]	0.002150	36.043937	0.544093	0.544200
MRCG	0.000005	0.728309	[−0.000023, 0.000032]	0.002066	2278.922227	0.540486	0.539771
MUVGn	0.000007	0.662562	[−0.000023, 0.000036]	0.002200	37.854180	0.499401	0.499340
RWEG	0.000015	0.212077	[−0.000009, 0.000039]	0.001780	207.669748	0.491927	0.491760
SAP	0.000012	0.325047	[−0.000012, 0.000037]	0.001832	130.001008	0.499455	0.499543
SDFG	0.000006	0.777292	[−0.000035, 0.000046]	0.003050	46.523493	0.528944	0.528663
SIEGn	0.000023	0.159366	[−0.000009, 0.000054]	0.002378	48.631774	0.516948	0.516768
SZGG	0.000017	0.444514	[−0.000026, 0.000059]	0.003214	36.311284	0.534831	0.534369
TKAG	0.000030	0.160174	[−0.000012, 0.000072]	0.003156	42.525102	0.523673	0.523504
VOWG	-0.000041	0.077097	[−0.000087, 0.000004]	0.003440	83.259032	0.538732	0.538921

3.4.3. ARMA-GARCH

By combining ARMA and GARCH models to an ARMA-GARCH model volatility clustering can explicitly be taken into account. An ARMA(1,1)-GARCH(1,1) model for the joint returns $R_t = \left(R_t^{(1)}, ..., R_t^{(n)} \right)$ at time t is formulated as follows

$$R_{t+1}^{(k)} = c_k + a_k R_t^{(k)} + b_k \sigma_{k,t} \eta_{k,t} + \sigma_{k,t+1} \eta_{k,t+1} \tag{3.4.1}$$

$$\sigma_{t,t+1}^2 = \alpha_{0,k} + \alpha_{1,k} \sigma_{k,t}^2 \eta_{k,t}^2 + \xi_k \sigma_{k,t}^2 \tag{3.4.2}$$

where $\eta_t = (\eta_{1,t}, ..., \eta_{n,t})^T$ is the multivariate standard NTS random vector with parameters $(\alpha, \theta, \beta, \rho)$. Consequently the return for each single asset is driven by a merely univariate ARMA(1,1)-GARCH(1,1) model. The only source of dependence for the different k is the consequence of the dependence structure of the vector of standardized innovations η_t, so the parameter ρ of the stdMNTS. The Portfolio Return is the weighted Sum of the different components:

$$R_{P,t} = \sum_{k=1}^{n} w_{k,t} T_{k,t}$$

$$w_t = (w_{1,t}, ..., w_{n,t}), \sum_{k=1}^{n} = 1$$

As the portfolio weights are equally, we further speak of the equally weighted portfolio (EWPF). Based on the transformed data we estimate different models for each stock an ARMA(1,1)-GARCH(1,1) model with Gaussian Innovation, an ARMA(1,1)-GARCH(1,1) model with NTS Innovation and an ARMA(1,1)-FIGARCH(1,d,1). After their seperate estimation we apply the multivariate case using the EWPF for the risk estimation and backtesting procedure. The models termed MNTS ARMA(1,1)-GARCH(1,1) and MNTS ARMA(1,1)-FIGARCH(1,d,1) employ the assumption of univariate t-distributions for heavy-tailed process innovations to efficiently determine estimates of the process coefficients. These estimates are subsequently used for generating the time series of standardized residuals. The standardized residuals of the 30 components are then further aggregated into the n-dimensional time series.

The estimates of the ARMA-GARCH NTS parameters for each of the 30 DAX stocks are given in table 3.3. Due to the shorter time series compared to the single index examination, we get different values for $\alpha = 1.3460$ and $\theta = 0.120$ values for the stdNTS distribution that are employed for the estimation of the remaining parameters for the 30 DAX stocks.

TABLE 3.3.: *ARMA-GARCH NTS Coefficients for DAX 30 Stocks*

RIC	c	a	b	α_0	α_1	η	β
ADSG	-0.000006	-0.746051	0.767355	0.000000	0.139974	0.786300	-0.007072
ALVG	0.000009	0.710146	-0.752578	0.000000	0.127779	0.831664	-0.031265
BASF	0.000007	0.746683	-0.775394	0.000000	0.128746	0.832960	-0.008421
BAYG	0.000007	0.746683	-0.775394	0.000000	0.128746	0.832960	-0.008421
BEIG	0.000001	0.901847	-0.919421	0.000000	0.159779	0.748388	0.011050
BMWG	0.000005	0.653991	-0.694293	0.000000	0.139583	0.801193	-0.000638
CBKG	0.000036	0.160263	-0.238737	0.000000	0.158198	0.817021	-0.047581
DAIGn	0.000009	0.642574	-0.691902	0.000000	0.149894	0.821049	-0.027858
DB1Gn	0.000020	0.267384	-0.351334	0.000000	0.147406	0.811023	-0.026563
DBKG	0.000016	0.465767	-0.491633	0.000000	0.137400	0.835154	-0.008450
DPWGn	0.000016	0.465767	-0.491633	0.000000	0.137400	0.835154	-0.008450
DTEGn	0.000005	0.424360	-0.500061	0.000000	0.132341	0.775367	0.002272
EONGn	0.000000	0.844946	-0.867733	0.000000	0.130624	0.803732	-0.034845
FMEG	0.000018	-0.647782	0.656302	0.000000	0.152488	0.771232	-0.003164
FREG	-0.000002	-0.102211	0.227339	0.000001	0.824487	0.175512	0.006037
HNKGp	0.000024	-0.178846	0.221144	0.000000	0.188974	0.678859	-0.005871
HNRGn	0.000006	0.328426	-0.337170	0.000000	0.136925	0.825116	0.001744
LHAG	0.000002	0.691803	-0.731529	0.000000	0.119667	0.821936	-0.006034
LING	0.000008	0.675776	-0.735289	0.000000	0.160517	0.751098	-0.008780
MANG	0.000006	0.394786	-0.429251	0.000000	0.163665	0.788194	-0.006636
MEOG	0.000002	0.827351	-0.854231	0.000000	0.138898	0.800028	-0.021949
MRCG	0.000003	0.819794	-0.853088	0.000000	0.206646	0.722347	-0.026343

TABLE 3.3.: *ARMA-GARCH NTS Coefficients for DAX 30 Stocks*

RIC	c	a	b	α_0	α_1	η	β
MUVGn	-0.000002	0.496600	-0.588563	0.000000	0.114877	0.818724	-0.021416
RWEG	-0.000001	0.623848	-0.655386	0.000000	0.129174	0.786597	-0.011231
SAP	0.000003	0.639752	-0.721627	0.000000	0.224371	0.664205	-0.018223
SDFG	0.000000	0.665357	-0.733658	0.000000	0.153818	0.785690	-0.013919
SIEGn	0.000009	0.507855	-0.557546	0.000000	0.133338	0.808178	-0.011266
SZGG	0.000007	0.233326	-0.293580	0.000000	0.177777	0.749945	-0.033771
TKAG	0.000007	0.496914	-0.544128	0.000000	0.138115	0.800576	-0.012803
VOWG	-0.000017	0.001812	-0.066460	0.000000	0.175632	0.804720	-0.022544

Followingly we compare the GoF test for the NTS ARMA-GARCH model and the AMRA-FIGARCH model with NTS and Normal innovation. Table 3.4 and 3.5 compare the achieved GoF values for the 30 DAX stocks. The NTS distribution proves superior in all scenarios over the Normal assumption. In case of the ARMA-GARCH NTS and ARMA-FIGARCH NTS no clear conclusion can be drawn on which model is superior.

TABLE 3.4.: *GoF tests for 30 DAX stock residuals of the NTS ARMA-GARCH model*

RIC	p-value	KS	AD	AD^2
ADSG	0.043999	0.016097	0.048813	0.000417
ALVG	0.002287	0.021452	0.068816	0.001216
BASF	0.029536	0.016918	0.046924	0.000854
BAYG	0.029536	0.016918	0.046924	0.000854
BEIG	0.060340	0.015416	0.052354	0.000598
BMWG	0.005342	0.020062	0.062114	0.000817

TABLE 3.4.: *Gof tests for 30 DAX stock residuals of the NTS ARMA-GARCH model*

RIC	p-value	KS	AD	AD^2
CBKG	0.000330	0.024325	0.064650	0.001348
DAIGn	0.005105	0.020139	0.413851	0.000638
DB1Gn	0.019486	0.017734	0.096634	0.000562
DBKG	0.000909	0.022868	0.067978	0.001219
DPWGn	0.000909	0.022868	0.067978	0.001219
DTEGn	0.000000	0.044449	0.099515	0.003037
EONGn	0.000000	0.032473	0.091141	0.002329
FMEG	0.000000	0.043186	0.086404	0.000855
FREG	0.000000	0.066207	0.135140	0.007730
HNKGp	0.000033	0.027348	0.063719	0.000471
HNRGn	0.000003	0.030373	0.060581	0.000431
LHAG	0.000000	0.054473	0.108945	0.002467
LING	0.051189	0.015774	0.047875	0.000385
MANG	0.024381	0.017299	0.048671	0.000515
MEOG	0.007621	0.019450	0.048930	0.000585
MRCG	0.108234	0.014068	0.074048	0.000449
MUVGn	0.000000	0.053865	0.118159	0.004014
RWEG	0.000179	0.025165	0.075814	0.001310
SAP	0.007613	0.019452	0.042679	0.000472
SDFG	0.000054	0.026734	0.075186	0.001372
SIEGn	0.000842	0.022981	0.066756	0.001289
SZGG	0.001140	0.022528	0.079577	0.001274
TKAG	0.003796	0.020633	0.064731	0.000987
VOWG	0.000000	0.034871	0.069176	0.000612

TABLE 3.5.: GoF tests for 30 DAX stock residuals of the ARMA-FIGARCH model with Normal an NTS innovations

RIC	Normal				NTS			
	p-value	KS	AD	AD²	p-value	KS	AD	AD²
ADSG	0.000000	0.057545	8.499885	0.007592	0.007504	0.019477	0.041348	0.000237
ALVG	0.000000	0.057545	Inf	Inf	0.002060	0.021617	0.067901	0.001033
BASF	0.000000	0.057545	Inf	Inf	0.057835	0.015509	0.041912	0.000351
BAYG	0.000000	0.057545	Inf	Inf	0.057835	0.015509	0.041912	0.000351
BEIG	0.000000	0.057545	Inf	Inf	0.047857	0.015919	0.047260	0.000401
BMWG	0.000000	0.057545	Inf	Inf	0.114465	0.013932	0.044499	0.000405
CBKG	0.000000	0.057545	Inf	Inf	0.000003	0.030033	0.068890	0.001421
DAIGn	0.000000	0.057545	Inf	Inf	0.006452	0.019739	0.721465	0.000603
DB1Gn	0.000000	0.057545	Inf	Inf	0.048929	0.015871	0.104425	0.000412
DBKG	0.000000	0.057545	Inf	Inf	0.001500	0.022110	0.066284	0.001237
DPWGn	0.000000	0.057545	Inf	Inf	0.001500	0.022110	0.066284	0.001237
DTEGn	0.000000	0.057545	126.526851	0.005409	0.000000	0.039078	0.089782	0.002522
EONGn	0.000000	0.057545	1367.864772	0.005370	0.000000	0.029567	0.082166	0.001794
FMEG	0.000000	0.057545	Inf	Inf	0.000083	0.026177	0.052224	0.000589
FREG	0.000000	0.057545	934.910267	0.023138	0.000000	0.061219	0.124727	0.006774
HNKGp	0.000000	0.057545	Inf	Inf	0.000005	0.029631	0.059697	0.000424

TABLE 3.5.: GoF tests for 30 DAX stock residuals of the ARMA-FIGARCH model with Normal an NTS innovations

RIC	Normal				NTS			
	p-value	KS	AD	AD^2	p-value	KS	AD	AD^2
HNRGn	0.000000	0.057545	12922.966452	0.006492	0.000000	0.033963	0.067922	0.000662
LHAG	0.000000	0.057545	99.417823	0.008428	0.000000	0.058765	0.117430	0.002504
LING	0.000000	0.057545	116.882727	0.007193	0.092695	0.014438	0.047336	0.000233
MANG	0.000000	0.057545	20.981716	0.006711	0.157326	0.013133	0.037712	0.000374
MEOG	0.000000	0.057545	8114.667878	0.007483	0.000000	0.032330	0.064668	0.000457
MRCG	0.000000	0.057545	Inf	Inf	0.119873	0.013819	0.100714	0.000408
MUVGn	0.000000	0.057545	54549.843418	0.008393	0.000000	0.060550	0.124208	0.004238
RWEG	0.000000	0.057545	52.453589	0.005341	0.000906	0.022873	0.065647	0.000896
SAP	0.000000	0.057545	5779.326292	0.006136	0.000713	0.023225	0.054717	0.000996
SDFG	0.000000	0.057545	1742.532441	0.005897	0.059784	0.015436	0.043083	0.000409
SIEGn	0.000000	0.057545	Inf	Inf	0.007423	0.019496	0.058432	0.000987
SZGG	0.000000	0.057545	20.061943	0.005407	0.003435	0.020797	0.069882	0.000983
TKAG	0.000000	0.057545	1482.365916	0.006064	0.005130	0.020130	0.051883	0.000462
VOWG	0.000000	0.057545	Inf	Inf	0.000000	0.034316	0.068131	0.000572

3.5. Comparison of Risk Measures for ARMA-GARCH Models

The purpose of this section is to show the significant difference in the distinctive model's performance on the application for estimating intraday risk

3.5.1. Risk Prediction

In order to adapt the model to currently available information in the best possible manner, one considers a time series model with continuous re-estimation. This increases the flexibility of the model as conditional mean and variance are estimated on times series models including all available information contained in past realizations. For the 30 DAX stocks we use 5 min high frequency data for the analyzed time period between April, 1 2009 and September, 1 2009 to show the VaR of the Normal ARMA(1,1)-GARCH(1,1) and the NTS ARMA(1,1)-GARCH(1,1) as well as the Normal ARMA(1,1)-FIGARCH(1,1) on September,1 2009.

TABLE 3.6.: *Risk Measures for different innovation distributions*

RIC	ARMA(1,1)-GARCH(1,1)			ARMA(1,1)-FIGARCH(1,d,1)		
	Normal	NTS		Normal	NTS	
	VaR	VaR	$AVaR$	VaR	VaR	$AVaR$
ADSG	0.005381	0.006108	0.008533	0.005161	0.007075	0.009842
ALVG	0.005610	0.007510	0.010592	0.005352	0.007508	0.010541
BASF	0.008298	0.011343	0.015864	0.006599	0.010457	0.014568
BAYG	0.008298	0.011343	0.015864	0.006599	0.010457	0.014568
BEIG	0.003354	0.003989	0.005588	0.003260	0.004088	0.005697
BMWG	0.005433	0.007102	0.009882	0.005364	0.007558	0.010473
CBKG	0.010378	0.012664	0.017857	0.009607	0.012696	0.017801
DAIGn	0.007982	0.010572	0.014855	0.007011	0.010518	0.014689
DB1Gn	0.007083	0.009688	0.013559	0.007281	0.009582	0.013337
DBKG	0.006216	0.007794	0.010911	0.005724	0.007632	0.010681
DPWGn	0.006216	0.007794	0.010911	0.005724	0.007632	0.010681

TABLE 3.6.: *Risk Measures for different innovation distributions*

	ARMA(1,1)-GARCH(1,1)			ARMA(1,1)-FIGARCH(1,d,1)		
	Normal	NTS		Normal	NTS	
RIC	VaR	VaR	$AVaR$	VaR	VaR	$AVaR$
DTEGn	0.004081	0.005130	0.007160	0.003644	0.004765	0.006622
EONGn	0.007987	0.009820	0.013845	0.006314	0.007814	0.010962
FMEG	0.003237	0.004076	0.005685	0.003089	0.004102	0.005708
FREG	0.004815	0.003750	0.005208	0.004552	0.007138	0.009880
HNKGp	0.005529	0.006521	0.009105	0.004838	0.006809	0.009474
HNRGn	0.004309	0.005418	0.007550	0.003926	0.005202	0.007211
LHAG	0.004707	0.005880	0.008243	0.004814	0.006434	0.008959
LING	0.004501	0.005917	0.008350	0.004258	0.006138	0.008623
MANG	0.005620	0.007434	0.010368	0.005673	0.008086	0.011221
MEOG	0.004466	0.005756	0.008097	0.003803	0.005505	0.007681
MRCG	0.003214	0.004091	0.005784	0.003635	0.004616	0.006474
MUVGn	0.004530	0.005752	0.008139	0.003753	0.004752	0.006694
RWEG	0.004533	0.005772	0.008077	0.003762	0.005170	0.007201
SAP	0.003886	0.005286	0.007434	0.003894	0.005365	0.007488
SDFG	0.005349	0.006560	0.009209	0.005416	0.007501	0.010484
SIEGn	0.005928	0.007773	0.010850	0.004735	0.006833	0.009490
SZGG	0.004861	0.005888	0.008266	0.005382	0.007330	0.010243
TKAG	0.006229	0.008056	0.011253	0.005388	0.008294	0.011543
VOWG	0.006070	0.007637	0.010689	0.005470	0.007987	0.011145

In Table 3.6 it is shown that the VaR using Normal innovation is generally lower than using NTS and that the difference between FIGARCH and GARCH is marginal. These results are to be further substantiated in section 3.5.2 to illustrate the behavior over time.

3.5.2. Backtestings

The various formal backtesting procedures described in section (2.5.5) concerning the adequat-
ness of an estimated time series model and its distributional assumptions for the implied standardized residuals are based on the number of quantile violations. The results of their different
variants are presented in this section. Specifically the MNTS ARMA-FIGARCH, the MNTS
ARMA-GARCH model and multivariate normal ARMA-GARCH model with dynamic parameter values are employed to generate distribution forecasts of EWPF returns for the time horizon
of 5 minutes. These dynamic forecasts are compared to the EWPF return realizations in a
backtesting period of 250 5 min time steps due to computational limitations.

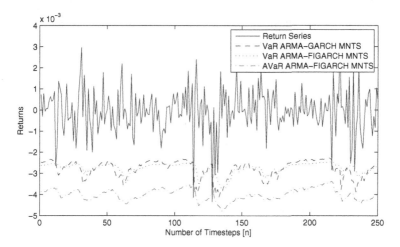

FIGURE 3.8.: *Backtesting of MNTS ARMA(1,1)-FIGARCH(1,d,1) and MNTS ARMA(1,1)-
GARCH(1,1) for backtesting window of 250.*

Figure 3.8 clearly displays the long-range dependence in the ARMA-FIGARCH model. But also
indicates a possible better fit of the MNTS ARMA-GARCH model, as due to the long-range
dependence the adjusting happens too slowly. The individual test variants are in case of the
Christoffersen test

- unconditional coverage (LR_{uc})

- test for serial independence (LR_{ind1})

- test for conditional coverage (LR_{cc})

and in case of the Berkowitz test

- lower tail test (LR_{tail})

- test for serial independence (LR_{ind2})

TABLE 3.7.: *Christoffersen Tests*

	Christoffersen tests		
Model	LR_{uc}(p-value)	LR_{ind1}(p-value)	LR_{cc}(p-value)
MNTS ARMA-GARCH	0.7691(0.3805)	0.1306(0.7178)	0.8998(0.6377)
MNTS ARMA-FIGARCH	3.5554(0.0594)	0.2963(0.5862)	3.8517(0.1458)

TABLE 3.8.: *Berkowitz Tets*

	Berkowitz tests	
Model	LR_{tail}(p-value)	LR_{ind2}(p-value)
MNTS ARMA-GARCH	1.2425(0.5373)	0.3897(0.5324)
MNTS ARMA-FIGARCH	0.1275(0.9382)	0.3721(0.5418)

4. Conclusion

One of the most pressing economic issues facing corporations today is the proper management of financial risks. Adequate statistical modeling of the underlying process is therefore essential. In this master thesis we investigate on the risk estimation on an intraday time scale of a five minutes interval on the DAX and its 30 stocks, as well as a multivariate approach in connection with long-range dependency. A multidimensional asset price process constitutes one of the principal tasks in several areas of modern quantitative

finance. One of the most critical aspects are asymmetry and features of heavy-tailedness of return distributions. Moreover, volatility clustering is observed on the time scale. These phenomena are widely acknowledged as stylized facts of

financial market data. After a brief introduction of financial econometrics and long range dependency, basic concepts and definitions of the theory of mulitvariate Lévy processes, the time-change of Lévy processes as well as a compact overview with respect of univariate tempered stable processes are reviewed. After the brief theoretical introduction the empirical analysis encompases a thorough investigation of the DAX30 index on an high frequency time-scale of 5 minutes.

In order to properly construct and apply conditional mean and variance models, the characteristic intraday volatility profile has been removed from the time series. After a detailed examination of the prevailing stylized facts and the existance of long-range dependence conditional mean and variance models have been applied, with a clear focus on ARMA(1,1)-GARCH(1,1) and ARMA(1,1)-FIGARCH(1,d,1) models. In the univariate as well as the mulitvariate case two innovation distributions have been investigated the normal and the Normal tempered stable distribution.

The assumption that the normal distribution is not capable of modeling the intraday innovation process was empirically proven in all tested cases. The NTS innovation assumption showed that in both approaches, the ARMA-GARCH and the ARMA-FIGARCH have better descriptive properties. The investigation of the long-range dependence led to the insight that long-range dependence is existing, but that the 5 min sampling period can be a case where long-range dependence is of less impact than for shorter time scales as found in the literature. This is supported by the results of a multivariate backtesting where it is shown that on a five minute sampling interval a ARMA-GARCH MNTS model represents the slighty better solution compared to an ARMA-FIGARCH MNTS model. Further research needs to be done in this area, especially on the less investigated DAX30 index.

A. DAX 30 Stocks

TABLE A.1.: *The 30 DAX stocks between 23.03.2009 and 21.09.2009*

Company	Industry	Member since	Headquater	RIC
Adidas	Clothing	22.06.1998	Herzogenaurach	ADSGn.DE
Allianz	Financial Services	01.07.1988	München	ALVG.DE
BASF	Chemicals	01.07.1988	Ludwigshafen	BASFn.DE
Bayer	Pharmaceuticals and Chemicals	01.07.1988	Leverkusen	BAYGn.DE
Beiersdorf	Consumer goods	22.12.2008	Hamburg	BEIG.DE
BMW	Manufacturing	01.07.1988	München	BMWG.DE
Commerzbank	Banking	01.07.1988	Frankfurt am Main	CBKG.DE
Daimler	Manufacturing	21.12.1998	Stuttgart	DAIGn.DE
Deutsche Bank	Banking	01.07.1988	Frankfurt am Main	DBKGn.DE
Deutsche Börse	Securities	23.12.2002	Frankfurt am Main	DB1Gn.DE
Deutsche Post	Communications	19.03.2001	Bonn	DPWGn.DE
Deutsche Telekom	Communications	18.11.1996	Bonn	DTEGn.DE
E.ON	Energy	19.06.2001	Düsseldorf	EONGn.DE
Fresenius Medical Care	Medical	20.09.1999	Hof an der Saale	FMEG.DE
Fresenius	Medical	23.12.2009	Bad Homburg vor der Höhe	FREG.DE
Hannover Rück	Financial Services	23.03.2009	Hannover	HNRGn.DE
Henkel	Consumer goods	01.07.1988	Düsseldorf	HNKGp.DE

TABLE A.1.: *The 30 DAX stocks between 23.03.2009 and 21.09.2009*

Company	Industry	Member since	Headquater	RIC
K+S	Chemicals	22.09.2008	Kassel	SDFGn.DE
Linde	Industrial gases	01.07.1988	München	LING.DE
Lufthansa	Transport Aviation	01.07.1988	Köln	LHAG.DE
MAN	Manufacturing	01.07.1988	München	MANG.DE
Merck	Pharmaceuticals and Chemicals	18.06.2007	Darmstadt	MRCG.DE
Metro	Retailing	22.07.1996	Düsseldorf	MEOG.DE
Munich Re	Financial Services	23.09.1996	München	MUVGn.DE
RWE	Energy	01.07.1988	Essen	RWEG.DE
Salzgitter	Steel	22.12.2008	Salzgitter	SZGG.DE
SAP	IT	18.09.1995	Walldorf (Baden)	SAPG.DE
Siemens	Industrial , electronics	01.07.1988	Berlinand München	SIEGn.DE
ThyssenKrupp	Industrial, manufacturing	25.03.1999	Essen	TKAG.DE
Volkswagen	Manufacturing	01.07.1988	Wolfsburg	VOWGp.DE

Bibliography

Aït-Sahalia, Y. and A. W. Lo (2000). Nonparametric risk management and implied risk aversion. *Journal of Econometrics 94*, 9–51.

Andersen, T. G. and T. Bollerslev (1997). Intraday periodicity and volatility persistence in financila markets. *Journal of Empirical Finance 4*, 115–158.

Artzner, P., F. Delbean, J.-M. Eber, and D. Heath (1998). Coherent measures of risk. *Mathematical Finance 9*(3), 203–228.

Baille, R. T., T. Bollerslev, and H. O. Mikkelsen (1996). Fractionally integrated generalized autoregressive conditional heteroskedasticity. *Journal of Econometrics 74*, 3–30.

Barndorff-Nielsen, O. E. and N. Shephard (2001). Normal tempered stable processes.

Beck, A. (2011). *Time Series Analysis and Market Microstructure Aspects on short Time Scales*. Ph. D. thesis, Karlsruhe Institute of Technology.

Beran, J. (1994). *Statistics for Long-Memory Processes*. Chapman & Hall.

Berkowitz, J. (2001). Testing density forecasts with applications to risk management. *Journal of Business and Econmic Statistics 19*, 465–474.

Bollerslev, T. (1986). Generalized autoregressive conditional heteroskedasticity. *Journal of Econmetrics 31*, 307–327.

Bollerslev, T., J. Litvinova, and G. Tauchen (2006). Leverage and volatility feedback effects in high-frequency data. *Journal of Financial Econometrics 4*, 354–384.

Bollerslev, T. and H. O. Mikkelsen (1996). Modeling and pricing long memory in stock market volatility. *Journal of Econometrics 73*, 151–184.

Boyarchenko, S. and S. Levendorskii (2000). Option pricing for truncated lévy processes. *International Journal for Theory and Applications in Finance 3*, 549–552.

Campbell, S. D. (2005). A review of backtesting and backtesting procedures. In *Technical Report 2005-21*, Federal Reserve staff working paper in the Finance and Econmics Discussion Series.

Carr, P., H. Geman, D. Madan, and M. Yor (2003). Stochastic Volatility for Lévy Processes. *Mathematical Finance 13*, 345–382.

Christoffersen, P. F. (1998). Evaluating interval forecasts. *International Economic Review 39*(4), 841–862.

Chung, C. (1999). Estimating the fractionally intergrated garch model. *National Taiwan University*.

Clark, P. (1973). A subordinated stochastic process model with finite variance for speculative prices. *Econometrica 41*, 135–155.

Conrad, C. and B. Haag (2006). Inequality constraints in the fractionally integrated garch model. *Journal of Financial Econometrics 4*, 413–449.

Cont, R. and P. Tankov (2004). *Financial modelling with jump processes*. Chapman & Hall/CRC.

Danielsson, J. (2011). *Financial Risk Forecasting*. Wiley.

Danielsson, J., B. N. Jorgensen, G. Samorodnitsky, M. Sarma, and C. G. de Vries (2010). Fat tails, var and subadditivity.

Ding, Z., C. W. Granger, and R. Engle (1993). A long memory property of stock market returns and a new model. *Journal of Empirical Finance 1*, 83–106.

Duffie, D. and J. Pan (1997). An overview of value at risk. *Journal of Derivatives 4*(3), 7–49.

Engle, R. F. (1982). Autoregresive conditional heteroskedasticity wiht estimates of the variance of u.k. inflation. *Econometrica 50*, 1287–1294.

Engle, R. F. and T. Bollerslev (1986). Modelling the persistence of conditional variances. *Econometric Reviews 5(1)*, 1–50.

Fama, E. F. (1965). The behaviour of stock market prices. *Journal of Business 38*, 34–105.

Fraenkle, J. (2010). *Theoretical and Practical Aspects of Algorithmic Trading*. Ph. D. thesis.

Francq, C. and J.-M. Zakoian (2010). *GARCH Models*. John Wiley & SonsLtd.

Greenspan, A. (2007). Symposium on maintaining financial stability in global economy.

Harris, L. (1986). A transaction data study of weekly and intradaily patterns in stock returns. *Journal of Financial Economics 16*, 99–117.

Hendershott, T. and R. Riordan (2012). Algorithmic Trading and the Market for Liquidity. *Journal of Financial and Quantitative Analysis forthcoming*.

Heston, S. (1993). A closed-form solution for options with stochastic volatility with applications to bond and currency options. *Review of Financial Studies 6*, 327–343.

Hull, J. (2007). *Risk Management and Finacial Institutions*. Pearson.

Hull, J. and A. White (1998). Incorporating volatility updating into the historical simulation method for value at risk. *Journal of Risk 1*, 5–19.

Itai, L. and A. Sussmann (2009). US equity high frequency trading: Strategies, sizing, and market structure.

Kim, Y. S., S. T. Rachev, M. L. Mitov, and F. J. Fabozzi (2011). Time series analysis for financial market meltdowns. *Journal of Banking and Finance 35*(8), 1879–1891.

Koponen, I. (1995). Analytic approach to the problem of convergence of truncated lévy flights towards the gaussian stochastic process. *Physical Review E 52*, 1197–1199.

Krause, D. (2011). *Portfolio Analysis with Mulitvariate Normal Tempered Stable Processes and Distributions*. Ph. D. thesis, Karlsruhe Institute of Technology.

Kupiec, P. (1995). Techniques for verifying the accuracy of risk managment models. *Journal of Derivatives 3*, 73–84.

Kyprianou, A. E. (2006). *Introductory lectures on fluctuations of Lévy Processes with applications*. Universitext. Springer, Berlin.

Linsmeier, T. J. and N. D. Pearson (2000). Value at risk. *Financial Analyst Journal 56*(2), 47–67.

Lopez, J. A. (1999). Methods for evaluating value-at-risk models. *Federal Reserve Bank of San Francisco Economic Review 2*, 3–17.

Mandelbrot, B. B. (1963). The variation of certain speculative prices. *Journal of Business 36*, 392–417.

Mandelbrot, B. B. (1967). The variation of some other speculative prices. *Journal of Business 40*, 393–413.

Mandelbrot, B. B. and M. S. Taqqu (1979). Robust R/S analysis of long-run serial correlation. In *Proceedings of hte 42nd Seesion of the International Statistical Institute* (Book 2 ed.), Volume 48.

Mandelbrot, B. B. and J. R. Wallis (1969). Computer experiments with fractional Gaussian noises, Parts 1,2,3. *Water Resources Research 5*, 228–267.

Menn, C. and S. T. Rachev (2009). Smoothly truncated stable distributions, garch models, and option pricing. *Mathematical Methods of Operations Research 63*, 411–438.

Merton, R. (1976). Option pricing when underlying stock returns are discontinuous. *Journal of Financial Economics 3*, 125–144.

Rachev, S., Y. Kim, M. Bianchi, and F. Fabozzi (2011). *Financial Models with Lévy Processes and Volatility Clustering*.

Rachev, S. T. and S. Mittnik (2000). *Stable Paretian models in finance*. Wiley.

Rachev, S. T., S. Mittnik, F. J. Fabozzi, S. M. Focardi, and T. Jašić (2007). *Financial Econometrics*. John Wiley & SonsLtd.

Riordan, R. and A. Storkenmaier (2012). Latency, liquidity and price discovery. *Journal of Financial Markets 15*(4), 416–437.

Samorodnitsky, G. and M. S. Taqqu (2000). *Stable non-Gaussian random processes*.

Sato, K.-I. (1999). *Lévy Processes and infinitly divisible distributions*. Cambridge University Press.

Taqqu, M. S., V. Teverovsky, and W. Willinger (1995). Estimators for long-range dependence: an empirical study. *Fractals. An Interdisciplinary Journal 3*, 785–798.

The Wall Street Journal (2007, August 11). One quant sees shakeout for the ages: 10,000 years. *The Wall Street Journal*.

Wood, R. A., T. H. McInish, and J. K. Ord (1985). An investigation of transaction data of NYSE stocks. *Journal of Finance 25*, 723–739.

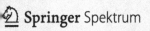